Engaging Indonesia

Critical Dialogues on Culture and Society

Engaging Indonesia brings together current debates in social and cultural studies relating to the study of Indonesia, with the aim of giving voice to both leading and emerging Indonesianists. Critical and interdisciplinary, the books in the series contribute to knowledge formation by addressing pressing issues facing the country—cultural, social, and political—from contemporary and historical perspectives. Works may include research-based monographs, edited volumes, handbooks, and translations of both recent, and seminal/historical works. The series is particularly welcoming of topics relating to issues of urban and environmental change, to questions of identity and place, to problems of governance and conflicts, to the politics of representation, language, and performances, and to the global and postcolonial contexts in which these are addressed. Special consideration is given to studies that cover underrepresented regions and marginalized voices, as well as those that seek to recover lost archives, uncounted political events, or forgotten historical figures. Relevant to students and scholars from a broad range of disciplines in the humanities and social sciences with a particular interest in researching Indonesia, the series champions intellectual engagement with the archipelago's rapidly changing landscape—from within, and beyond.

Melani Budianta · Manneke Budiman ·
Kathrin Oester Znoj
Editors

A Liveable Kampung

The Challenges of Urban Expansion in Greater Jakarta and East Nusa Tenggara

Springer

Editors
Melani Budianta
Universitas Indonesia
Depok, Indonesia

Manneke Budiman
Universitas Indonesia
Depok, Indonesia

Kathrin Oester Znoj
Institute of Social Anthropology
University of Bern
Bern, Switzerland

ISSN 2948-2909 ISSN 2948-2917 (electronic)
Engaging Indonesia
ISBN 978-981-96-7865-5 ISBN 978-981-96-7866-2 (eBook)
https://doi.org/10.1007/978-981-96-7866-2

The project was supported by Pinter-Midi Grant, Directorate of Research and Development, Universitas Indonesia.

This Springer imprint is published by the registered company Springer Nature Singapore Pte Ltd.
The registered company address is: 152 Beach Road, #21-01/04 Gateway East, Singapore 189721, Singapore

Acknowledgments

This edited volume would not have been possible in the first place without an instrumental support from Prof. Rosari Saleh, then Vice-Rector for Research and Community Engagement of Universitas Indonesia, who generously provided a special research grant for an interdisciplinary research project in the university through the creation of a research scheme PINTERMIDI UI (the Indonesian acronym for Inter- and Multi-Disciplinary Research of Universitas Indonesia) in 2019. Not only did this program provide a multi-year research funding, but it also allowed—perhaps for the first time—the meeting of researchers from ten faculties and schools within the university as well as researchers from several partner institutions outside the university that occurred in an intense and productive way, and thus breaking the disciplinary walls that for a long time had prevented a collaborative work at such a scale to take place.

Special thanks also go to postgraduate students, early-career researchers, and consultants whose contributions to the research had been invaluable and without whose help the core research team of the PINTERMIDI would have had to deal with many daunting methodological and technical obstacles. Some of them also contributed to the writing of the chapters in this volume, and their names are listed as contributors in the various chapters.

We also convey our gratitude to Prof. Diah Ayu Maharani, then the Head of the Research and Innovation Management Office of Universitas Indonesia, who facilitated the early preparation for the publication of this book. In conjunction with this, our gratitude also go to Dr. Shuri Mariasih Gietty Tambunan, then Vice-Dean for Academic, Research, and Student Affairs of the Faculty of Humanities, and Ms. Pratidina Sekar Pembayun from the Faculty of Humanities' publication team, for their constant and all-out support in ensuring that this book is eventually published by establishing contact and maintaining communication with Springer Nature.

Last but not least, we cannot appreciate enough the solid support, patience, and understanding extended to us by Ms. Alex Westcott Campbell and her team in Springer Nature. Their assistance has been simply incredible.

To all of them, we are forever indebted.

Manneke Budiman
Melani Budianta
Kathrin Oester

Contents

List of Figures

List of Tables

Part I

Chapter 1
Introduction: Liveable Kampungs: Challenges in Urban Expansion

Manneke Budiman, Melani Budianta, and Kathrin Oester

Abstract This introductory chapter presents an overview of the existing studies on kampungs in the context of urbanization that invades what used to be rural areas as a result of urban development, particularly in developing countries like Indonesia. It also provides the highlight of each chapter of this edited volume and key arguments proposed by the contributors, which demonstrate the ways in which the research findings of the chapters offer an array of novel ideas in regards to the nature of relationship and interaction between the urban and the rural, which enrich further the existing contemporary studies of urban–rural dynamics, especially in Asian urban settings. Furthermore, this introduction also hints at how the chapters explore how the communities in the urban peripheries create and mobilize their local "commonings" to help them develop resilience in the face of the unstoppable urban expansion.

Keywords Urban expansion · Commoning · Urban–rural dynamics · Urban studies · Community resilience

In 2013, in a chain of alleys in the back of an industrial estate in Tangerang City, southwest of Jakarta, stood rows of small houses in impoverished conditions and unsanitary environments. Bottled inside the houses are families of five or six, with children playing in the alleys filled with litter. The sight was considered common among informal, unplanned settlements called kampung, inhabited by lower-class families, that spread all over Jakarta. In 2014, then the Governor of Jakarta, Joko Widodo announced a program called "thematic kampung", whereby each kampung community was encouraged to adopt one theme that should relate to a collective project they

M. Budiman · M. Budianta (✉)
Department of Literature, Faculty of Humanities, Universitas Indonesia, Depok, Indonesia
e-mail: melani.budianta@ui.ac.id

M. Budiman
e-mail: manneke.budiman@ui.ac.id

K. Oester
Institute of Social Anthropology, University of Bern, Laenggassstrasse 49, Bern, Switzerland
e-mail: katharina.oester@faculty.unibe.ch

M. Budianta et al. (eds.), *A Liveable Kampung*, Engaging Indonesia,
https://doi.org/10.1007/978-981-96-7866-2_1

could do in their kampung, which could transform their living environment to be a livable one.

Inspired by this thematic kampung project, Suli Rosidy, then the head of a sub-district in the Tangerang City, to the West of Jakarta, listed around 30 most unhealthy kampungs in his administration and challenged them to make a change within 6 months through a program called Clean and Healthy Living (*Program Hidup Bersih dan Sehat*). There was no subsidy for the program, but the winner would receive some amount of prize money and be honored in a public event (Budianta 2019).

Within 6 months, changes occurred in the neighborhoods mentioned in the list, as kampung communities started to clean up their environments. Not all of the winners could make the change sustainable, but those who did managed it through socially engineered collective action. Among these efforts are creating collective urban farming, establishing kampung waste management and clean water supply systems, and clearing up alley corners for better common space.

In the 1990s, the existence of kampungs as "unregulated popular sectors" of housing in Jakarta was to be predicted to be eliminated by the government policy toward regulated housing in Jakarta's move from a big village to a modern city (Leaf 1992). Although the policy for promoting private housing, business, and industrial areas has cleared up most of the greater Jakarta areas from unregulated popular sectors through voluntary or forced eviction and gentrification, the kampungs hung on tenaciously in pockets in between spaces. The interdependence of formal and informal economic sectors, the need for affordable housing near job opportunities for low-income wage earners, and the steady influx of people migrating from the countryside are among the causes of the continuing need for urban kampung to exist in megacities like Jakarta (Hellman et al. 2018). Kusno called this kampung as a "leftover space" that are "often off the official map…but "are located at the 'heart' and on the 'margins' of planned middle-class neighborhoods" and has its own ways to survive urban expansion (2023, p. 64).

Based on the 2019 statistics, the greater metropolitan area of Jakarta plus its surrounding peri-urban areas (Jakarta, Bogor, Tangerang, Depok, Bekasi) include a population of around 30 million people on 4384 sq kms. In the Jakarta city area alone, the population density is 16,704/km. As the metropolis is more and more expanding, it swallows surrounding areas, with a population growth rate of 3% (World Population Review 2021).

Outside Jakarta and other big cities in Java and other islands, *desakota*, villages with economies connected to the cities, are also urbanized at a fast pace. According to the United Nations Department of Economic and Social Affairs, Population Division (2018), the world urban population will reach 68% in 2050, and 90% of the urban growth will occur in Asia and Africa. Indonesia is among the Asian countries with the fastest rate of urbanization (Lewis 2013). As Boras (2009) noted, globalization and neoliberal capital have affected not only agrarian production and the economy as a whole with its working conditions, but it also further made borders between rural and urban areas porous. In many rural areas in Indonesia, the expanding market of agricultural commodities, such as oil palm and cacao, are responsible for a decreasing subsistence economy, including rice production.

Some studies have also observed that the consumption of fast food and processed food are widespread, even among youth in rural areas (Sharkey et al. 2011). At the same time, Boras (2009) underlines that land grabbing, urban migration, and unequal distribution and access to power and capital have rendered rural areas vulnerable. The inequality between urban centers and rural peripheries is exacerbated by imbalanced access to resources and an economic gap between Western and Eastern Indonesia. One obvious evidence of such a gap is the prevalence of stunting in Eastern Indonesia (Laksono and Wulandari 2023; Titaley et al. 2019).

Facing urban expansion with its gains and consequences, urban kampungs in Tangerang, and villages in Nangapanda and Flores struggle to create their own livability in their own ways. The "urban kampung" and the "urbanized village" are two types of vulnerable communities, both of which rooted in urban expansion. Having in mind the sustainability of future cities, the improvement of the kampungs' precarious conditions is imperative. Urban kampungs, located at the intersection of formal and informal economies, provide a constant influx of flexible workers; cities are therefore challenged to support them by improving their living conditions.

Similarly, urbanized villages due to their human and agricultural resources are an important part of the general supply chain. The two kinds of kampungs, within and outside the cities, are also connected through the cycle of migration from the rural to the urban area and vice-versa. How are these two types of kampungs afflicted by urban expansion and what role do they play when planning a sustainable future? These are the main questions raised in this edited volume.

1.1 Aspects of Liveability: An Interdisciplinary Perspective

In a macro-analysis of developmentalism, urban kampungs are usually sidelined as problems to deal with, and not as entities to be valued as they are. The existence of kampungs is one of the characteristics of "Third-World" cities that used to be colonial cities in the Southeast Asian region (McGee 1971, qtd in Rimmer and Dick 2019, p. 10), and these cities, as Rimmer (1977) argued, were engulfed one by one by the "accelerating process of globalization" (qtd in Rimmer and Dick 2019, p. 10). These cities are, among others, Singapore, Kuala Lumpur, Metro Manila, Bangkok, and Jakarta.

In the surrounding areas of these cities, villages were replaced by new manufacturing and industrial estates, which in turn created a rising demand for new housing areas for the urban middle-class that also grew along the urban expansion to the peripheral areas (Dick 1990, and Rimmer 1986, qtd in Rimmer and Dick 2019, p. 12). As a result, according to Rimmer and Dick, these cities failed to provide adequate urban infrastructure and have been facing a logistic crisis (2019, p. 12). Land transport system is an inseparable part of the mega-expansion of a metropolitan region like Jakarta (McGee 2002, p. 647). The construction of toll roads around the capital city has brought about a radical change to the surrounding areas of the cities and new towns. The first toll road, Jagorawi, began to be built in 1973 and was opened in 1978 connecting Jakarta, Bogor, and Ciawi. This toll road helped create *desakota*, a label used by Terry McGee in describing the expansion of the mega-region of Jakarta. It is a mix of urban and rural setups and lifestyles that puts kampung around and nearby residential and industrial areas (McGee 1991, quoted in Kusno 2023, p. 55).

The land conversion that massively took place for the sake of urban use in this context of urban expansion, most often, was not accompanied by adequate "planning or coordination" (Yap 2019, p. 31). An urban region, such as the Greater Metropolitan Area of Jakarta, nowadays includes large cities (the five sub-regions of the capital city of Jakarta), small cities (such as Depok, Tangerang, Tangerang Selatan, and Bekasi) and small towns (such as Bintaro, Serpong, Ciputat, Karawaci, Cileungsi, Cibubur), villages and rural areas (such as Kampung Nambo, Kampung Markisa, Desa Cimone), which—in Yap's term—"displays a mixture of urban and rural features" (2019, p. 31).

With reference to Jakarta, we need to add another feature, namely, urban kampungs, such as Kampung Cikini and Gedong Pompa, located within the capital city (Central Jakarta and North Jakarta) and strongly display such an urban–rural mixture. Yap pointed out that what is happening now is the new urban areas "increasingly encircling rural areas" rather than the other way around (2019, p. 32). This phenomenon is called "peri-urbanization" (Webster 2011, qtd in Yap 2019), which results in the emergence of a type of "fragmented landscape" that is neither urban nor rural. Kusno stated that in between these "unplanned" landscapes, emerged "*jalan tikus*" (literally means "rat-pathways") or "para-transportation systems" that both "separate and connect communities" in which motorbikes, minivans, and bicycles transport workers from kampung to the surrounding new residential estates and industrial complexes (2023, p. 55).

In the hinterland, rural areas, remote from the capital city, experience urbanization as well because of decentralization. More careful thought is needed to understand the social transformation that takes place as a result. This is especially relevant when thinking about what is happening in a rural region, such as Nangapanda in Flores. Tania Murray Li (1999) stated that every research on the hinterland (such as the villages in Nangapanda) has the potential of offering breakthroughs, insofar as we consider issues of marginalization (between the upland and the lowland), power and its manifestations, and the forms of production that are based on local political, economic, and environmental conditions.

Li further pointed out that there is no singular model of transformation of life that takes place in rural communities in the hinterland, as some areas may become more densely populated because of urbanization and industrialization, while some other areas may witness a decrease in population for urbanization and environmental destruction (1999). That is why, the approach we used in this book regarding the Nangapanda region in the island of Flores is significantly different from those used to discuss the peri-urban and urban kampung areas within and around the capital city of Jakarta, with the focus being more on health infrastructure issues than water supply, sanitation, and waste management.

Microanalytic studies of kampungs, especially from a socio-cultural perspective, abound: they show people's resilience and coping strategies (Budianta and Hapsarani 2019) but also failed government strategies and precarious living conditions. Poorly planned peri-urbanization in big cities like Jakarta has a devastating impact on the natural environment within and surrounding the city. Yap stated that both legal and illegal waste disposal is "probably the earliest form of exploitation of the peri-urban zone by the city" (2019, p. 36). New housing and industrial estates do not always have wastewater treatment facilities, and wastewater is simply channeled into rivers or directly to the sea (such as in Gedong Pompa, North Jakarta, which does not have a piped water supply or sewer system).

Pollution has plagued the peripheral areas of big cities in the form of mercury leaks, hazardous waste, and harmful odors, such as what has happened in one of the sites researched in this book, Kampung Nambo, in South Tangerang. According to Kundu (2007), the population inhabiting the peri-urban areas (or urban kampungs, for that matter) probably has a low-level awareness of the environmental hazards resulting from industrialization processes and, thus, is less prepared in dealing with the arrival of polluting industries (qtd in Yap 2019, p. 37).

In most cases, even local administrations are overwhelmed by the rapid and massive urban and peri-urban transformation because they have to deal with challenges that are brought by both "local and global market forces" (Yap 2019, p. 37). In Indonesia, as Yap found, decentralization has caused local administrations to be more "inward-looking, egocentric, and uncooperative", and respect for plans and regulations are low (2019, pp. 37–38). Decentralization also puts new pressure on local governments to be more creative and "entrepreneurial" in mobilizing local resources, but more often than not, this has been translated into converting more land for urban use without being supported by an adequate capacity to "plan and manage urban development" (Yap 2019, p. 38).

Human capital is the most fundamental factor in understanding how urban expansion can be managed and its destructive impact minimized. Forbes (2019) suggested that knowledge and creativity are what need to be fostered among the population. They are "interconnected" and "overlapping", as knowledge is associated more with "formal education", whereas creativity is often the "product of tradition and imagination" that are not directly related to formal education (Forbes 2019, p. 45). Forbes argued that demographic change in the Southeast Asian region creates an increasing demand for education (2019, p. 51), whereas Budianta (2019) suggested that rural and kampung communities possess "cultural commonings" (*lumbung budaya*), which serve as resources of strength, resilience, and sustainability. They can also be learning resources for creative ways to come to terms with urban expansion and the new challenges brought about by the forces of urbanization.

Community-based approaches to urbanization issues in the heart of the city, as well as in the peripheral areas, should be used as a basis for working toward alternatives in which the collective effort of intra-communities and inter-communities could be encouraged (Bunell et al. 2019), especially in light of the fact that city authorities and local administrations do not always have the capacity to provide the necessary services and resources, even if they want to (Boonyabancha 2009, qtd in Bunell et al. 2019, p. 56). In the early years of the New Order, Ali Sadikin, the legendary governor of Jakarta (1966–1975), tried to "clean up" the city by imposing order to its citizens. He viewed seasonal workers from the rural areas with a low level of education and poor skills, such as pedicab drivers, construction workers, street-vendors, homeless people, beggars, and sex workers as a critical problem for public order. Sadikin associated, these poor urban dwellers with kampung, which—to him—implied "backwardness" (Kusno 2007, p. 36–7).

This edited volume focuses on the kampungs' livability through an interdisciplinary approach. In such research, one discipline's perspective can be enriched and modified by another, resulting in a deeper understanding of complex processes and structures. As an example, one kampung's preliminary survey gave promising insights into the collective enterprise of urban farming; however, when examined by different parameters its waste management showed high levels of toxicity endangering public health. As the example shows, studying livability by an interdisciplinary approach opens up unexpected perspectives.

This edited volume is the result of a series of meetings carried out by researchers from various disciplines, namely the faculty of medicine, public health, technology, sociology, psychology, humanities, and economics. They conducted research in various kampungs within and outside Jakarta. Revisiting the results, a team of multidisciplinary researchers reexamined the issues of WASH (water, sanitation, health) and socio-economic capital.

This volume's contributions cover five kampungs in the Greater Jakarta area, namely Cimone, Markisa, and Nambo, located in Tangerang City and South Tangerang, southwest of Jakarta, Cikini in Central Jakarta, and Gedung Pompa in North Jakarta. All five kampungs in Jakarta have a long history of dealing with gradual urban expansion, they are struggling with socio-economic precarity, with issues in water supply, sanitation, housing, and public health.

To give a broader view, it seemed important to contrast the capital's perspective in a second section of the volume with perspectives from Eastern Indonesia. The contributions focus on an area located in the Nangapanda sub-district in the vicinity of Ende, Flores. Here the issue of livability centers on nutrition, stunting, and education—reflecting the gap in equity and access. One final chapter about an urban kampung in Surabaya gives a different dimension not yet covered by the book on the contribution of Chinese-ethnic actors.

1.2 Chapters Outline

The edited volume consists of two sections, the first one, including six contributions, addresses the kampungs in the Greater Jakarta area, and the second one, including three contributions, the kampungs in Nangapanda District, in Ende, Flores.

Chapter 2 examines what livability means in two kampungs from a sociological perspective (Estu Putri Wilujeng) and the perspective of architecture (Hanifah Fijriah). The two authors examined the way the two urban kampungs are dealing with basic service problems concerning water, sanitation, and hygiene. They collected their data with an integrated method and concluded that the two kampungs, Kampung Cikini Kramat in Central Jakarta and Kampung Gedung Pompa in North Jakarta, using social and cultural networking strategies as well as socio-spatial tactics to deal with the complexities of urban expansion.

Chapter 3 approaches WASH (water, sanitation, health) conditions from a socio-economic perspective. Written by Zakianis and Rita Damayanti (public health), Arief Wibisono Lubis (economics), and Jenny Anggita (independent researcher), the chapter examines five kampungs in the Greater Jakarta Metropolitan Area. Collecting data from 717 households, this research explored the ability of the urban poor to pay for garbage transportation and collection, but also their willingness to pay for wastewater management. The findings reveal that the typical/average costs for a household's drinking water and sanitation services were less than or equal to 4% of the minimum wage in the region. For local government policymakers, this type of economic and WTP-willingness to pay (WTP) mapping is crucial when deciding on appropriate tariffs for WASH management in urban kampungs.

As a way of dealing with the rising amount of domestic waste due to consumption growth, kampung communities often start to produce compost, which is used in urban farming. In this way, a circular economy that generates sustainable green living is created. This sounds like the best strategy to create a livable kampung. However, Chapter 4, written by Gabriel Andari Kristanto, M. Adhiraga Pratama, Ihsan Ramadhan, and Rusyda Maheswara (engineering), critically challenged the urban composting solution. They emphasized, instead, that contaminated organic waste can cause major health risks when used to grow vegetables in urban farming. In four kampung areas in Greater Jakarta, the study examined the waste's heavy metal content in an 8-day survey. The authors found that the compost produced from organic waste caused significant health risks for the population. To minimize them, especially contamination of heavy metals, proper waste management is necessary as well as continuing education on waste separation.

Chapter 5 raises the issue of air quality in kampungs that are located near landfill sites. Ova Candra Dewi and Gabriel Andari Kristanto (engineering), Zakianis (public health), Nadia Adrianti and Intan Chairunnisa (medicine), investigated the microbial air quality in Kampung Nambo, South Tangerang. With its steep topography, Kampung Nambo is located less than 300 m from Cipeucang landfill. The landfill has been operating since 2011 and received municipal solid waste from Tangerang Selatan City. The waste decomposition produces odors and germs that spread all over Kampung Nambo and beyond. About 74 houses out of 153 have been tested for their microbial air quality with the result that 67% exceeded the permitted limits.

Measuring the houses' orientation, elevation, cross ventilation, and distance from the landfill, the researchers generated a four-quadrant matrix that allowed them to determine and visualize different air polluting factors. Outdoor and indoor air flow was simulated through ANSYS, which allowed the visualization of air movement. As a result, the study found that cross ventilation is not easily possible in houses adjacent to a landfill due to the bad outdoor air quality. In addition, the authors developed workable solutions to enhance the air quality for households living under unfavorable conditions.

The next two chapters are complementary as each of them deals with the use of herbal medicine in the kampungs of greater Jakarta from a different perspective. While Chapter 6 focuses on the use of herbal medicines, Chapter 7 tackles the issue of self-medication, especially the mixture of herbal medicines and prescription drugs.

In Chapter 6, the authors—Manneke Budiman, Bantari Wardhani (medicine), Halvina Saiya (environment), Dhita Hapsarani (humanities), and Rista Ihwanny (independent researcher)—aimed at examining the local communities' resilience through using herbal medicines for self-medication despite modernization and urbanization in Jakarta and Tangerang. While most scholars in the field discussed the persistence of herbal medicines in urban communities from a general point of view, only a few if any, look deeper into the socio-political motives that underlie such persistence.

Using a combination of ethnographic methods and questionnaires, the authors found a correlation between the persistence of herbal medicine in urban and peri-urban communities and the specific way they deal with modernity brought by urbanization. Part of it is to cope with the policies and regulations concerning the utilization of herbal medicines implemented by health authorities. As the authors argued, such negotiations are essential for the communities' survival and the persistence of their local, social, and cultural identities. Thus, while they cannot put a halt to the invading force of modernity, they still attempt to leave some room for their traditions and identities.

Chapter 7 complements Chapter 6 by focusing on the practice of self-medication in urban kampungs. Based on a pre-tested structured questionnaire, the authors discussed the results of a cross-sectional study in 150 households in Gedong Pompa Kampong, North Jakarta. The study includes demographic data and practices and perceptions of self-medication. The authors, Bantari Wardhani, Indira Rahmawati, Juwita Sari and Nabila (medicine), and Melva Louisa (pharmacology), started with the hypothesis that, if done responsibly and in accordance with health standards, urban kampung dwellers' self-medication could be an effective strategy to reduce their healthcare expenses. Recently, however, an accelerated flow of information through the internet and electronic media, especially in urban kampungs of Greater Jakarta, led to the potential misuse of certain prescription drugs.

The authors found that 18.7% of the respondents admitted to using a mixture of drugs (they call it "*obat rombongan*") obtained from nearby stalls for all kinds of health problems. Upon further interviews, they found out that the mixture of drugs consisted of vitamins, NSAIDs, antibiotics, and corticosteroids. In spite of such inappropriate practices, most respondents did have sufficient knowledge regarding self-medication. 60% of the respondents affirmed self-medication when sick. A total of 47% stated that the drugs used in SM were safe. Most respondents also agreed that it was mandatory to ask for a doctor's advice before self-medication. This contradiction between knowledge and practice is a research gap to be closed with future studies on survival strategies in urban kampungs.

The second section opens with Chapter 8, aiming at assessing the sustainability of villages in Nangapanda District (Kecamatan Nangapanda), Flores, East Nusa Tenggara Province, Indonesia. Gabriel Andari Kristanto (engineering), Halvina (economics), Dini Kemala (sociology), and Sanca Pamungkas (independent researcher) conducted a multi-dimensional survey collecting data on demographics, education, as well as health, and nutrition in 347 households. Part of the survey were 1130 children, 995 from 24 elementary schools and 175 aged 2–4 years. Furthermore, 21 water samples from springs, reservoirs, and taps in Nangapanda sub-district were examined regarding physical, chemical, and biological parameters. As a result, the authors identified various social, health, and environmental aspects that hindered Nangapanda District on its way to developing sustainable village environments. Furthermore, they offered recommendations of how to deal with these challenges in rural Nangapanda District.

The next two chapters, Chapter 9 and Chapter 10, complement one another by discussing different aspects of nutrition and stunting in East Nusa Tenggara. Chapter 9 was written by Raphaella Dewantari Dwianto, Putri Ayu Irmaya (sociology), Grace Wange, and Dini Suciyanti (nutrition studies). As the authors argued, eating—including nutritious food—is not just a type of behavior but a social practice. East Nusa Tenggara was chosen for its high prevalence of stunting—the highest in Indonesia. In 2019, three villages, namely Ndoturea, Ndororea, Ndororea 1, were selected for data collection with quantitative as well as qualitative methods (participant observation). In order to measure the nutrition level, the survey included 1000 children under age 5 and at primary school age. Randomly, the study chose 200 parents to get an overview of the family's eating practices.

As a result, the study confirmed the prevalence of stunting in one district in East Nusa Tenggara, namely Nangapanda. Here, stunting happened despite the parents' knowledge on nutritious food and their local knowledge of how to prepare food. The gap between relevant knowledge on nutrition and the parents' social practices is closely linked to their economic situation, which affected the family's eating practices in negative ways.

Chapter 10 extends the discussion on stunting in Nangapanda, East Nusa Tenggara, arguing that a key issue is education strengthening children's cognitive skills. Three scholars from Psychology, Sr. Retno Pudjiati, Dhisty Azlia Firnady, and Rizqina Permatasari Ardiwijaya, argued that there is a close relationship between stunting and the parents' inability to give their children optimum cognitive stimulation. The study is based on 188 pairs of 5–9 years old children and their parents residing in rural Nangapanda. As a result, the study showed that children's learning experiences, their parents' education level, and the family's community participation affected children's cognitive development. The study's findings serve as a basis for future interventions aiming at the empowerment of families and teachers.

The last chapter of the book, Chapter 11, consists of an essay that imposes a historicized racial dimension on the space called kampung in Indonesia, with examples from urban kampungs in Jakarta and Surabaya. The chapter also highlights another aspect not discussed earlier in the preceding chapters, i.e., the involvement of activists in advocating for kampung sustainability and the threat of eviction initiated by local government authorities. The Epilogue sums up the findings of the chapters by mapping the kampung issues in Indonesia to date, considering the diversity as well as the similar patterns from western to eastern Indonesia. Reflecting on the livability of kampungs in Indonesia, the epilogue raises questions about the kampung's future in Indonesia given the pace of the expanding urbanism.

1.3 Afterthoughts

This book is a compilation of chapters that attempt to envision Indonesia from a different angle than the mainstream construction, in the sense that it tries to gain a better understanding about urbanization as a global phenomenon and its impact on communities at the local level. Thus, Indonesia is viewed from the peripheries, and so is the way in which it is (re)constructed. The contributors of this book also share a belief that an inter- and multidisciplinary approach is indispensable for that effort, not in order to build a macro-analysis of each of the sites that are researched but to tease up the complex nature and the problems that each site is facing, with the ultimate objective being to show that there is no singular explanation that can satisfactorily cover all of the diverse issues related to urbanization in the local–global context in Indonesia today. At the same time, the variety of geographic examples, different topics, and disciplinary approaches, collected in this volume, also point to similar difficulties in urban and semi-urban kampungs: while a globalized market economy has easy access to the kampungs' workforce, little attention is paid by companies and administrations to ecological questions and people's health. It is precisely here that some kampungs develop creative solutions, thereby drawing on hitherto hidden cultural resources.

References

Budianta, M. 2019. Smart kampung: Doing cultural studies in the global south. *Communication and Critical/cultural Studies* 16 (3): 241–256. https://doi.org/10.1080/14791420.2019.1650194.

Budianta, M., and D. Hapsarani, eds. 2019. *Meniti Arus Lokal–Global: Jejaring Budaya Kampung [Following local–global currents: Village cultural network]*. New York: Infermia Publishing.

Bunell, T., D. P. S. Goh, and H. Ng. 2019. Aspirations in urban(izing) southeast Asia. In *Routledge handbook of urbanization in southeast Asia*, ed. R. Padawangi, 54–63. New York: Routledge.

Boras, S. M., Jr. 2009. Agrarian change and peasant studies: Changes, continuities and challenges—An introduction. *The Journal of Peasant Studies* 36 (1): 5–31. https://doi.org/10.1080/03066150902820297.

Dick, H. W. 1990. Further reflections on the middle-class. In *The politics of the middle class in Indonesia*, ed. R. Tanter and K. Young, 63–70. New York: Centre for Asian Studies Monash University.

Forbes, D. 2019. Knowledge, creativity and the city. In *Routledge handbook of urbanization in southeast Asia*, ed. R. Padawangi, 43–53. New York: Routledge.

Hellman, J., M. Thynell, and R. van Voorst. 2018. *Jakarta: Claiming spaces and rights in the city.* New York: Routledge.

Kusno, A. 2007. *Penjaga Memori: Gardu di Perkotaan Jawa.* London: Penerbit Ombak.

Kusno, A. 2023. *Jakarta, the city of a thousand dimensions.* London: NUS Press.

Leaf, M. 1992. *Land regulation and housing development in Jakarta, Indonesia: From the "big village" to the "modern city".* University of California, Berkeley. ProQuest Dissertations Publishing, 1992. 9304978.

Lewis, B. 2013. Urbanization and economic growth in Indonesia: Good news, bad news and (possible) local government mitigation. *Regional Studies* 48 (1): 192–207.

Li, T. M., ed. 1999. *Transforming the Indonesian uplands: Marginality, power and production.* Berlin: Harwood Academic Publishers.

McGee, T. 1991. The emergence of desakota regions in Asia: Expanding a hypothesis. In *The extended metropolis: Settlement transition in Asia*, ed. N. Ginsburg, B. Koppel, and T. McGee. London: University of Hawaii Press.

McGee, T. 2002. Jalan, jalan: Invading, destroying, and reconstructing the Southeast Asian city. *Bijdragen Tot der Taal-, Land- En Volkenkunde* 158 (4): 673–752.

Rimmer, P. J. 1977. A conceptual framework for examining urban and regional transport needs in Southeast Asia. *Pacific Viewpoint* 18 (2): 133–148.

Rimmer, P. J., and H. Dick. 2019. Gateways, corridors, and peripheries. In *Routledge handbook of urbanization in southeast Asia*, ed. R. Padawangi, 9–30. London: Routledge.

Sharkey, J. R., C. M. Johnson, and W. R. Dean, et al. 2011. Focusing on fast food restaurants alone underestimates the relationship between neighborhood deprivation and exposure to fast food in a large rural area. *Nutrition Journal.* https://doi.org/10.1186/1475-2891-10-10.

World Population Review. 2021. *Jakarta Population Review 2021.* https://worldpopulationreview.com/world-cities/jakarta-population

Yap, K. S. 2019. Peri-urban transformations in Southeast Asia. In *Routledge handbook of urbanization in southeast Asia,* ed. R. Padawangi, 31–42. Routledge.

Factors Related to Stunted Toddlers in Eastern Indonesia

Titaley, C. R., I. Ariawan, D. Hapsari, A. Muasyaroh, and M. J. Dibley, et al. 2019. Determinants of the stunting of children under two years old in Indonesia: A multilevel analysis of the 2013 Indonesia basic health survey. *Nutrients* 11 (5): 1106. https://doi.org/10.3390/nu11051106

World Urbanization Prospects. 2018. *United Nations, Department of Economic and Social Affairs, Population Division* 16.2, accession 2018.

Part II

Chapter 2
The Future of Urban Kampung: Discourse Contestation, WASH Basic Services, and Social and Cultural Capital

Estu Putri Wilujeng and Hanifa Fijriah

Abstract This chapter examines the sustainability of urban kampung in the future. From the perspective of modernization, an urban kampung is known as a problematic, late-blooming area associated with slums. Their status has also generated problems that have led to inhabitants consistently facing potential eviction. On the other hand, urban kampung holds its history and cultural identities passed down through generations. Amid these situations, urban kampung inhabitants strive to fulfill their basic needs, which include water, sanitation, and hygiene (WASH). This research applied mixed methods and triangulated quantitative data from the 2019 general survey we did on WASH conditions and qualitative data on the process of the kampung strategies to have better WASH access. A causal-layered analysis was conducted to understand how to transform the present condition into the urban kampung alternative future. The two urban kampungs for our research are Gedong Pompa in the Jakarta coastal area and Cikini Kramat in Central Jakarta. The findings show that the two kampungs have developed everyday tactics to expand their networks in the form of reproduction and exchange of social and cultural capital to the other related actors. Nevertheless, they can still maximize their social and cultural capital for a better WASH system in the long-term planning of their WASH commons.

Keywords Cultural capital · Everyday tactics · Future studies · Urban kampung · Social capital · WASH system

---

E. P. Wilujeng (✉)
Department of Sociology, Faculty of Social and Political Sciences, Universitas Indonesia, Depok, Indonesia
e-mail: estuputri@ui.ac.id

H. Fijriah
Achitecture Program, Faculty of Engineering and Design, Institut Teknologi & Bisnis Ahmad Dahlan, South Tangerang, Indonesia
e-mail: hanifafijriah@itb-ad.ac.id

M. Budianta et al. (eds.), *A Liveable Kampung*, Engaging Indonesia,
https://doi.org/10.1007/978-981-96-7866-2_2

2.1 Introduction

Urban kampung is a contemporary social phenomenon that has existed since the colonial era. The existence of urban kampung in cities can be observed throughout Southeast Asia, such as Indonesia and Malaysia. While the term urban kampung is often conflated with the ghetto and the suburban, which are also scattered in cities. There are crucial differences between them. The main difference between urban kampung and these marginal urban spaces lies in the structure and cultural dimensions which can be seen in their everyday life.

The continual existence of urban kampung depends on the actors with different interests and discourse competition. Some scholars consider urban kampung as slums left behind in the march to modernity, while others view them as the cultural buffer zones of cities like Jakarta and have claims to the city (Kusno 2015, 2020). Amid debates about the value and role of urban kampung in cities, the residents of urban kampung continue to struggle on a daily basis to meet basic needs, such as water, sanitation, and hygiene (e.g., Nursyahbani and Pigawati 2015; Riogiang 2016).

Amid both pressure and support over space and issues of basic needs in the urban kampung, we argue that residents of urban kampung develop everyday tactics to maintain their living space and fulfill their basic needs. Everyday tactics are defined as a series of practical and strategic ways that are carried out daily to achieve certain goals. These tactics are part of the process of capital accumulation and exchange, both social and cultural capital. This is done so that they can survive and solve the more immediate problems of meeting basic needs, such as water, sanitation, and hygiene (WASH).

The problem in meeting WASH needs and the contention about space ownership are observable in two urban kampungs with different characteristics in the Special Capital Region of Jakarta, namely Gedong Pompa and Cikini Kramat. Gedong Pompa is a kampung in North Jakarta that is close to the coastal area, while Cikini Kramat is located in Central Jakarta. The differences in spatial characteristics and problems faced by these two kampungs implicate the tactics they practice in maintaining space and simultaneously meeting WASH needs.

The everyday tactics in maintaining space and meeting WASH needs can be understood as the process of exchanging existing capital with access to space and needs from various parties. Bourdieu (1986) classifies capital into three types, namely economic, cultural, and social. For Bourdieu, economic capital is the driving force for cultural and social capital. As such, residents of urban kampung who generally belong to middle- and lower-income groups are still able to gradually maintain space and meet the necessities of life. This resonates with Ostrom (1990) and his argument that communities have their internal model of management to achieve goals beyond the conditions set by external parties. The management of the commons is essentially the community's tactic of living as a collective. This collective action can also be managed in terms of social capital (Putnam in Ahn and Ostrom 2008). Lumaksono (2013) examined individual and collective strategies for the residents of Kampung

Jomblang when facing a water crisis by setting up water outlets, subscribing to gallons and wells, and conducting water treatment and filtering.

Unfortunately, urban kampung does not have a stronger economy, social, and cultural capital than the other communities. Moreover, they tend to be marginalized. This situation has raised questions about the possibility of urban kampung's future, especially when we compare it with other urban communities. As we argue, the future of urban kampung depends on the social and cultural capital that they have accumulated in their everyday tactics.

To understand the possibility of urban kampung's future, we study the history of the urban kampung, including the various existing regulations. What kind of problems are faced by urban kampung? What are their everyday tactics in terms of the reproduction and exchange of cultural and social capital? Moreover, what are the implications for its future? In order to answer these questions, we employed a mixed research method by combining data from the 2019 PINTERMIDI Survey to determine demographic characteristics, problems of WASH, and social capital with qualitative data from observations of daily life, interviews, and various supporting documents.

2.2 The Future of Urban Community: Literature Review

2.2.1 Socio-spatial Perspectives on Urban Space

Urban space, which is vital in a global context, often becomes the target of prophetic claims made by urban researchers and practitioners. Various paradigms have been employed to study urban spaces in the past, present, or future, including ecology, political economy, and the socio-spatial paradigm. Gottdiener (2012) noted a competition between human ecology and political economy. By examining these two paradigms, Gottdiener attempted to reconstruct the socio-spatial paradigm. The socio-spatial paradigm assumes that space can affect all kinds of components in an urban setting at various levels, from regional, national, and international. Spaces are also formed from the capital of multiple stakeholders which can be observed from the everyday. For this reason, we elaborate on the political-economic and socio-spatial approaches to studying the future of cities.

Gottdiener (2012) described a socio-spatial approach developed from a study of the everyday in urban and suburban communities as part of a larger metropolitan area. Urban and suburban spatial arrangements have manifest and latent consequences that affect individual behavior. This arrangement and its implementation are influenced by government policies, investors, and various real estate institutions that regulate individual behavior, preferences, and consumption. On the other hand, consumption and other individual behavior can also change spatial arrangements and lead to new spaces that are more in line with people's needs and desires. In this case, space holds a dual character as a place for action and social relations. This dual relation between society and space is manifested in various cultural symbols and material objects in everyday life.

This perspective can be used to describe urban spaces and communities, particularly how communities and spatial spaces interact with each other in everyday life. The relationship between community and space is also influenced by various groups which play a role in the spatial tower of urban communities, specifically in the interactional model that is carried out. These various components carry out a two-way interaction to maintain and reproduce the existing spaces.

2.2.2 Urban Future and Community: Commons and Socio-cultural Capital in Everyday Tactics

Understanding the future of cities cannot be separated from historical transitions and the role of the communities. Gottdiener (2012) wrote that from 10,000 years ago there has been a process of urbanization or the construction of dense buildings and areas. From a socio-spatial perspective, ancient cities are the product of power, economic function, and belief expressed in various symbols which together form an integrated system of meaning.

Today, the city has been developed with various symbols as part of envisioning the future on a global level. Nevertheless, this development process is often trapped to follow the other cities' models even though they have different characteristics. Urban development is also often trapped to reproduce towering and colossal buildings without considering the historical context and the characteristics of the space. For example, in Asia, which has a history of rural areas with its own set of problems, urban development models from Australia and the United States are championed, despite their problems, for instance, the city model of Los Angeles has been abandoned in urban planning. Cities in Asia are forced to follow these models as a global economic driver, for example, the development of Petronas in Malaysia, but the result is the opposite of what these models claim (Inayatullah 2011). Research shows a correlation between skyscraper buildings and the onset of an economic downturn (Asia Pacific Management News, 1999 in Inayatullah 2011).

Inayatullah (2011) further argued that the future of cities is expected to focus on overall health, both physical and social. This argument diverges from the assumption of the apocalyptic future of the city, such as the destruction caused by divine wrath and nuclear war; all of which causes city residents to fear their involvement in the survival of the city, including the assumption that the city will be destroyed by the presence of new communication technology. However, what happened is the opposite, since the process of providing internet access still relies on physical space (Gottdiener 2012).

Urban life, which is far from apocalyptic schemes and the impasse due to the emergence of the internet, shows a glimmer of hope for the future of the city. At times, cities plan various alternatives for their future. Inayatullah (2011) suggested that some cities, such as Seattle and Brisbane, have tried to portray some of their urban alternative futures. These cities invest in urban planning that can reduce the impact of climate change. They not only invest in building roads, rates, and rubbish but also develop environmental-friendly energy systems, such as transportation options. Cook (2018) stated that future studies always place the concept of climate change as the basis for future community problems. However, the problems faced by individuals in a broad range of possibilities for the future are not only related to climate change.

Pacione (1990) mapped two types of problems that will be faced by cities in the future. The first is the consequences of urban size and a population density which impact the urban environment, including urban crime rate. The second is the internal spatial structure of the city related to the distribution of welfare services and physical structures, along with their impact on transportation and mobility and their overall influence on interactions within the city. This deepens the segregation of urban space. Inayatullah (2011) described the urban problem as the deepening of segregation in frenetic individual progress in obtaining access to various services.

Amid these problems, city residents, especially those who are marginalized, develop a series of tactics carried out on an everyday basis. For example, Wutich (2011) examined the problem of water scarcity in the Chocabamba region, Bolivia. He also argued that in meeting water needs in conditions of scarcity, these residents exchange water based on the principle of reciprocity as part of the moral economy of water. Unfortunately, the principle is not a solution for residents who experience water scarcity due to political and economic pressures that cause resource gaps and social exclusion among the urban poor. Round et al. (2008) stated that tactics and strategies in informal groups are complex, as in the case of post-Soviet Ukraine. They argued that the strategies and tactics reflect the power relations and scale of power between the formal and informal economies. This is also carried out by informal workers and can be applied in residential spaces with power relations. Among the informal, formal, and unplanned-formal settlement areas, there are power relations that determine the struggle for space and services.

The life of urban poor and marginalized groups are subject to discriminatory experiences despite their status as fellow citizens of the city. This is caused by inequality in capital ownership. Bourdieu (1986) defined capital as accumulated labor which is attached to ideas and bodies, also termed as embodied form, and matter. Capital can be a source of social energy in the form of concrete and living labor (reified or living labor) which benefits the owner. In its accumulation, capital takes time.

Bourdieu (1986) mapped three types of capital, namely economic, cultural, and social. Economic capital is the basis of cultural and social capital. Economic capital can be converted into money and institutionalized in the form of property rights. Meanwhile, cultural capital is manifested in three forms, starting from the embodied state, namely the embodiment of the mind and body, the objectified state, namely the physical form of cultural objects (cultural goods), and the institutionalized state which can be seen from educational qualifications. Meanwhile, social capital is defined as an aggregate of potential and actual resources that is manifested in a durable form of ownership of a network or relationship between two or more parties that mutually acknowledge each other. These mutual relational networks can be institutionalized.

The size of social capital is calculated from the size of the network that is connected and can be mobilized as well as the volume of capital owned by the network linked to interested parties.

The concept of social capital can be traced to a study developed by Louis Hanifan in 1916 but under different terms, namely good faith, fellowship, mutual sympathy, and social intercourse in forming social units carried out by groups of individuals and families (in Aldrich and Meyer 2015). Aldrich and Meyer (2015) also defined social capital and its use in community resilience before, during, and after a disaster. They follow the classification of social capital types, namely bonding, bridging, and linking (Aldrich 2012; Kawachi et al. 2004; Szreter and Woolcock 2004 in Aldrich and Meyer 2015). Bonding social capital is defined as a relationship between individuals who have emotional closeness, usually within the scope of family and friendship. Meanwhile, bridging is defined as partners or individual relationships within a broader social group, such as class and race, as well as various organizations that can help individuals by providing information and resources that are useful in surviving in society. Finally, linking social capital is the relationship between individuals and groups that have power that is based on a sense of trust.

Bourdieu (1986) noted that social capital can be reproduced using continuous socialization and exchange which will result in recognition and can be reproduced. The reproduction process can be carried out both in internal relations and in external relations of communities in urban areas. Rolalisasi et al. (2013) further stated that social capital can assist communities in urban areas to secure a better life.

Several studies also demonstrated the relationship between community members and other parties, such as NGOs and the state, in improving residential areas. Winayanti and Lang (2004) described how residents residing in informal settlements practice self-help for certain problems, such as self-help for improving the quality of housing through independent financing and development. In addition, the involvement of NGOs and the government helps increase access to services for communities in informal settlement areas, such as when the state puts pressure on kampung to hasten it in applying for RT/RW legality and when NGOs assist residents of Kampung Penas Tanggul to avoid eviction through amassing various sources.

This situation demonstrates that the success of a kampung to survive and meet its rights is the result of a combination of independent efforts and how they build relationships between residents and external institutions outside the kampung and actors with more power. If executed well, this will benefit residents in the future.

2.2.3 Kampung as Part of the Urban Community

The concept of kampung has a wide range of definitions, starting from settlements in villages to cities. Bunnell (2002) described kampung in Malaysia as a rural settlement that has historically undergone urbanization. Bunnell argued that from a social perspective, kampung in Malaysia is an urban space. The process of urbanization is not only a movement toward urban physical space but also a process of adopting the values of urban life. Bunnell also stated that the kampung is socially, culturally, and economically integrated with the national and global political economy. The kampung concept used by Bunnell (2002) refers to rural areas undergoing transition. Similar to the Indonesian context, the word kampung originally refers to the character of the countryside with its close social relations, mutual assistance, communalism, and intensive communication. This character can be observed in urban areas which then led to the concept of urban kampung. Devas (1981) defined kampung as an informal, unplanned area, without public services that can be observed in urban areas. Based on a study of resilience in urban kampung in Surabaya, Shirleyana et al. (2018) contended that the traditional kampung model is the dominant residential model in Indonesia which contradicts the development model of megacities.

Both urbanized kampung and urban kampung with their rural characteristics obscures the boundaries of the kampung-city dichotomy. Both are also equally influential in national and global political-economic relations. Shirleyana et al. (2018) argued that kampungs are resilient in urban life; they have the potential to support urban livability and livelihoods by providing affordable housing and other informal economic activities. The type of housing they provide becomes an option for old occupants or newcomers who are studying, working, or getting health services in the megacity district.

The transition of kampung and its history into a lower-class area began during the colonial period. During that period, the definition of kampung was reproduced from village and kampung settlements inhabited by Non-Europeans and Non-Chinese in urban areas. The European and Chinese were considered upper-class groups, so this resulted in kampung becoming an area of middle and lower-class groups. The implication was that there was spatial segregation based on national ethnicity and social class.

The transformation of kampung into a location for the lower class made it increasingly difficult to obtain urban rights. The living space is constantly under threat of eviction; the shortage of clean water holds the potential to disrupt the existence of urban kampung. Yuwono and Wardiningsih (2016) argued that the existence of urban kampung needs to be maintained by optimizing internal and external values. This will address the problem of urban kampung which is the fact that it is deemed incompatible with urban modernization, and this is identical to the kampung space model. One of them is in the Kuningan area which was evicted in the 1990s despite its historical and cultural values, namely the myths and legends of Prince Kuningan, the mosque, and his legacy. The prince's tomb is said to have been transformed into a magnificent building in the Kuningan area. Other historical remains in the Kuningan area that still exist are the Al-Mughni tomb and mosque. For this reason, they recommended a mechanism for developing internal–external relations in the kampung by preserving the kampung, developing its tourism potential, developing historical, social, and cultural values, and carrying out integrated regional development, including the involvement of residents.

Urban kampung also has its strength as a lifeline for city life. Kusno (2015) argued that Jakarta's strength is formed by the interaction of three main components, namely the creative destruction of capitalism, the violence of state categories, and kampung as the force of the vernacular environment. In the midst of state and capital, villages hold the potential of human agencies who network with each other to survive socially and economically. Their way of survival is that they can continue living under the pressure from the state and capital while still meeting the basic needs of the residents. Kampung becomes a capital space for residents, who are usually migrants working in the formal and informal sectors.

2.3 Methods

To understand the possibilities of urban kampung's future, the authors map the past and present of urban kampung's issues, and their everyday tactics to increase their cultural and social capital by using mixed methods. The authors triangulated quantitative and qualitative data. The quantitative data were obtained from the 2019 PINTERMIDI Survey. The qualitative data have been gathered from the interview, focus group discussion, and related documents such as the regulation and the latest research on urban kampungs like Cikini Kramat and Gedong Pompa.

The 2019 PINTERMIDI Survey is an interdisciplinary survey on sanitation issues, such as water and air, as well as access to lighting in Greater Jakarta urban kampung. We take 300 samples from the raw data based on the location of the respondents, like in Cikini Kramat and Gedong Pompa. We take the variables on social capital, like group and network memberships, moral trust, and society. For the group and network membership, we take some items, such as the participation of members in groups in the neighborhood, the participation of members in groups outside the neighborhood, the number of close friends, the number of people outside the family who are willing to help if the respondent loses regular income, the number of people who have come to the respondent for help with personal problems in the last 12 months, the possibility the respondent relies on a neighbor to take care of children if the respondent is away for 1–2 days. For the moral trust and solidarity, we took several items, like respondents' general trust in the residents where the respondent lives, the possibility that the residents where the respondent lives will be willing to help when the respondent is in need, the possibility that the residents where the respondent lives are willing to lend money when the respondent needs them, informants the respondent can trust.

Besides the survey, we also used interview and observation data to understand the issues and everyday tactics or urban kampung. The authors analyzed the discourse of urban kampung development based on various regulations and research, both on intervention and identification. Then, we studied the level of social capital of the two villages, both Cikini Kramat and Gedong Pompa, with data also from the 2019 PINTERMIDI Survey Data as well as interviews and observations.

The data were processed and used as material for analysis of opportunities for future descriptions of urban kampung using the causal-layered analysis approach developed by Inayatullah (2008, 2013). In a causal-layered analysis, the analysis process is classified into four stages. First, the litany is a description of the existing problem situation and its implications. In this case, the team will use the data from the mapping of problems that occur in Cikini Kramat and Gedong Pompa. The second stage is systemic, which is the identification of social, economic, and political causes for the problems of meeting the basic needs of WASH in these two urban kampungs. The third stage is the worldview or cultural view of global relations. Fourth, the myth or metaphor contains various myths about urban existence, especially among urban kampung communities. After carrying out the four levels of analysis, the writing team constructed the possibilities that would occur in the urban village and the process of fulfilling expectations.

2.4 Looking at the Past of Urban Kampung in Jakarta: Transformation and Regulation

According to Silver (2008), kampung is defined as an urban version of a village in the countryside. Kampung usually consists of a group of dense single-story buildings (sometimes two stories or more) that are closely located, are connected by a network of paths, and do not have adequate sanitation. The meaning of kampung is literally 'camp' and is usually applied only to groups of native settlements. But in the mind of the Dutch during the colonial period, kampung meant lower-class urban settlement with the following attributes: lack of modern facilities such as water, sewers, and electricity; land governed by traditional or adaptation law; buildings that rely on informal construction methods (mostly using bamboo); high-density development; and the collapse of home and work. In addition, during the Dutch colonial period, kampung communities were often distinguished based on race and ethnicity rather than class, for example, Chinese kampung, Arab kampung, and so forth. The poor kampung residents are considered to have no geopolitical power at all. This is reinforced by evidence suggesting that at that time wealthy *pribumi* or natives bought land from their poorer neighbors in anticipation of future land use changes.

The existence of urban kampung in Jakarta is inseparable from the development of the population during the colonial period. Initially, the population of Batavia in the nineteenth century grew slowly until 1870 when the arrival of Europeans to Java increased rapidly. Between 1900 and 1930, the city's population increased from 115,000 to 435,000, representing an average annual growth rate of 9%. Five years later, through the annexation of the municipality of Meester Cornelis (now Jatinegara), Batavia's population jumped to more than half a million. Most of the urban population growth during this period was generated by immigration from rural areas and abroad which was accommodated by the conversion of agricultural areas to urban settlements, as well as densification in various native kampungs.

Batavia was then prepared to meet the infrastructure needs of a growing European housing community. In 1918, Batavia established the Batavia Regional Drinking Water Company (*Water Leidingen Bedriff van Batavia*) to manage a piped water distribution system that supplies 89 L per second of clean water from nine artesian wells in the Ciomas-Ciburial-Bogor area to European and other foreign communities in Batavia. The local water utility then expanded the piped water system to several kampungs, especially after the government agreed to partially finance a limited kampung improvement program that began in 1927 on the condition that Europeans received 4 times as much water as the natives. However, the housing conditions in urban kampung were deteriorating at that time. Nas in Silver stated that Europeans were increasingly aware of the dangers to their own health caused by unsanitary conditions in kampung located within their city limits but politically outside the jurisdiction of the city. Evictions and relocation were carried out to establish new cities, such as Menteng. The transformation of urban kampung continues to this day.

Marcussen in Silver (2008) suggested that there are three distinct periods of kampung transformation in Jakarta, the first of which began in the early 1900s when many areas were 'cleared' to pave the way for new urban development (such as Menteng). It was also during this period that the problem of sanitation and services for kampung residents were first given serious attention. The second phase took place after Independence when most of the villages were included in the new Jakarta administration. During this period, the kampung area increased in size and density to accommodate mass migration from rural areas. The third period was after 1965 when kampung was integrated into the New Order's comprehensive administrative and socio-political system. During this period a massive kampung improvement program was undertaken to provide better services to urban residents.

One program that is celebrated and considered successful is the Kampung Improvement Program or KIP. The program was launched systematically during Governor Ali Sadikin's administration which sought to redevelop Jakarta into a modern city. During this period, kampung covered about 60% of the land area of the city of Jakarta and accommodated about three-quarters of the population. According to a 1969 survey of conditions in about 500 villages in Jakarta, only a few had access to basic services (particularly WASH). 68% of houses in kampung did not have private toilets; 90% had no tap water; 80% had no access to electricity; many villages located along canals/rivers were used as garbage dumps; and there were relatively few permanent housing units. Only 24% of the buildings had solid walls, floors, and roofs, while almost 44% had structures made of bamboo and thatched roofs, and the other 32% were a combination of temporary and permanent (semi-permanent). KIP was not intended to address structural shortages in housing but to focus on environmental improvement through minimum infrastructure improvements that will encourage residents to maintain the environment they live in. 'Roads, drainage gutters, communal water taps, toilets, trash bins, and social services, such as schools and health clinics,' were tools used to improve the quality of the living environment and prevent a relocation that can disrupt community life (Silver 2008).

According to Silver (2008), the fourth stage in the development of kampung that could be added to the Marcussen periodization took place from the late 1980s to the year 2000. This was a period when the scale of kampung evictions across the metropolitan area increased as a direct result of aggressive private redevelopment, which was a part of the process to become a modern metropolis.

The implementation of KIP was divided into three phases, namely phase 1 in 1969–1974 with a total improved area reaching 2000 ha, phase 2 in 1974–1976 with a total improved area reaching 1980 ha, and phase 3 in 1976–1979 with a total improved area reaching 3000 ha. By the time the Jakarta KIP was completed in the late 1980s after covering most urban kampung areas, the government began embracing alternative strategies to meet the growing demand for more focused housing initiatives. The urban development strategy for Jakarta in the 1980s and 1990s then focused on a new private housing concept by partnering with PERUMNAS, which mostly operated in remote areas where land prices were cheaper. With this strategy, most of the kampung KIP redevelopment could even be relocated to build new neighborhoods and expand commercial areas in modern metropolitan cities. However, the high cost of property in the city made it difficult to operate this strategy in the kampung area in Jakarta (Silver 2008).

According to Silver (2008), new kampung continued to emerge in 1989 but their development shifted from the inside to the outside of the city. This is due to the clearing of a large area in the city center which had housed kampung for decades for new development. The widespread relocation of kampung residents became one of the most contentious issues of the late 1980s and 1990s, particularly the question of how much compensation should be paid to those evicted. This condition then led to a new strategy where the evicted kampung residents were relocated to a new multi-story building (now known as vertical housing or *Rumah Susun*). Citizens were convinced that this was not a mere ‘eviction program’ but ‘urban renewal; or urban renewal. In other words, even though all the inhabitants were evicted, the construction of the flat provides a guarantee that this activity was not just eviction.

Kampung Gedong Pompa and Cikini Kramat are two of the many crowded kampungs in Jakarta that also experienced this dynamic. Kampung Gedong Pompa was originally a swampy area inhabited by approximately 10 families. Some of the Gedong Pompa areas experienced eviction and relocation to vertical housing after the eviction of the Pluit Reservoir around 2014. Kampung Cikini Kramat itself has existed since the Dutch colonial period. Based on a map published by the Dutch in 1825, the area around Kampung Cikini has been referred to as ‘Kg. Tjiekienie.’ The early conditions of Cikini Kramat are not yet known, but many residents have lived for three generations in the area. In addition to KIP, these two kampungs have also been registered in the Community Action Plan or CAP program since 2018 which was initiated by Governor Anies Baswedan. The program is an empowerment program for the urban poor as part of a form of housing restructuring against forced evictions that occurred in slum areas in Jakarta. Slightly different from KIP, according to Muhtadi (2019), CAP puts active community participation as a goal and not just an approach (Muhtadi 2019). CAP, which collaborates with several stakeholders, also focuses on the quality of kampung settlement areas by ensuring residence and meeting the rights of Jakarta residents in terms of social, economic, and environmental aspects (Muhtadi 2019).

2.5 Urban Kampung Jakarta Today

2.5.1 The Problems of Urban Kampung in Jakarta: Cikini Kramat and Gedong Pompa

The existence of urban kampung in Jakarta cannot be generalized. Each area has certain history and characteristics. The team mapped out two urban kampung areas with different characteristics, namely Cikini Kramat and Gedong Pompa. Those two kampungs are both located in the Special Capital Region of Jakarta but have different geographical conditions and spatial characteristics. Simply put, Cikini Kramat is in the center of Jakarta, while Gedong Pompa is located in a coastal area in North Jakarta. However, this space position difference influences the demographic characteristics and problems related to space and WASH.

2.5.1.1 Gedong Pompa

Gedong Pompa is one of the kampungs in the administrative area of RT 002 and RW 017 in the Penjaringan subdistrict. RT or *Rukun Tetangga* means neighborhood association and RW or *Rukun Warga* is community association. According to the informants, RT 002 has 805 families with a population of around 1500 people. This neighborhood has been divided into three kampungs, Kampung Gedong Pompa, Kampung Nelayan, and Kampung Kebon Pisang or Bonpis. Kampung Gedong Pompa is the largest kampung in this neighborhood. It also became the 'capital' of RT 02. The division of kampung into RTs is interestingly based on the identity or material symbol attached to each area. The name of *Gedong Pompa*, or 'water pump building' can be traced to the existence of a water pump building in the area. *Kebon Pisang*, or literally 'banana plantation' can also be traced to the area which in the past was a site for banana plantation. *Nelayan* or literally 'fishermen' is located closest to the coast where many fishermen work and live.

Gedong Pompa residents who work as fishermen are fewer than those who work as fishermen in Kampung Nelayan. Some work by looking for fish and shrimp, and some also charter boats. Some of the kampung residents also work in urban industries as entrepreneurs or employees.

The kampung residential area is very dense. The population of the RT is larger when compared to most RT in Indonesia. In this kampung, there are many rented rooms and rented houses, so in one house there can be three or more families. There is no distance between houses, and they are separated by narrow alleys. The residents are not only physically close to each other, but they also have relatively strong social relations. Some residents even have shared kitchens.

Regarding WASH, residents have two public toilets and two types of water sources, namely a well (without charge) and a tap without a meter (with charge; people usually call it a tank). Both sources were claimed to be undrinkable due to contamination of seawater and waste, but tap water is considered cleaner. Therefore, for consumption purposes, most residents choose to buy refilled/bottled water.

Gedong Pompa residents use both water from the well and the tap for bathing. For the first rinse, residents use water from the well and for the second and last rinse, after using soap, they use tap water to make sure the soap suds are truly rinsed off. Many residents also have access to private clean water. Respondents claim that there is a crisis of trust among residents regarding the use of clean water. With many residents living in rented rooms or rented houses that do not provide private sanitary facilities, many also lock access to private tap water in their respective houses to avoid water theft. This is because the use of water has to be paid for. Therefore, clean water sources are protected so that they are not overused.

The kampung's geographic location, which is close to the coast of Jakarta, makes it frequently flooded due to the rise in sea level every 3–5 years. According to respondents, flooding is occurring more frequently since the land reclamation project. Even under regular conditions, residents can no longer use groundwater due to seawater intrusion and pollution from adjacent septic tanks and wells. The project thus makes it more difficult for residents to access clean water and to carry out other sanitation activities.

In addition to geographical challenges and limited land, the kampung also faces geopolitical issues. The term eviction has become increasingly familiar to residents. Like most densely populated kampung in Jakarta, changes to Jakarta's spatial plan are an issue to be tackled together as a collective. In this case, the residents of Gedong Pompa cooperated with various parties such as other kampung residents experiencing the same fate. They also joined UPC (Urban Poor Consortium) and JRMK (Urban Poor People's Network) as well as collaborated with academics and NGOs (RUJAK—center for urban studies) to discuss with government representatives.

2.5.1.2 Cikini Kramat

Cikini Kramat is a *Rukun Warga* 01 (community association) in the subdistrict of Pegangsaan, Menteng, Central Jakarta. It is divided into 13 *Rukun Tetangga* (RT or neighborhood association). Each neighborhood association is inhabited by around fifty to sixty families. This is different with Gedong Pompa, if the Kampung Gedong Pompa is part of a neighborhood association, Kampung Cikini Kramat is a whole community association, but their inhabitants are different. The number of neighborhood associations between those areas is also different. Then, the urban kampung classification does not depend on the urban administration with the exact indicators.

The classification of urban kampung can be traced to the historical context and the space material symbol. Based on the history of their space, Kampung Cikini has existed since the Dutch colonial period. Nowadays, this kampung still exists and has become a strategic area in Jakarta, Indonesia's capital city. The kampung is surrounded by major buildings such as ministry offices, hospitals, central supermarkets, and so forth. For this reason, many kampung residents work as employees and entrepreneurs.

Since the beginning, the kampung has been an important buffer zone for the surrounding area. Initially, many kampung residents worked as service providers for upscale housing in Menteng. On the other hand, since 1967, after the tram line was dismantled, the kampung seems to have been split in two by the existence of an elongated traditional market located on the ex-tram line. Until now, the market still exists and has even become a source of income for the kampung residents. Currently, Cikini Kramat also plays an important role for urban communities in providing necessities such as rented rooms, rented houses, and affordable food. Its strategic location which is close to the Cikini Commuter Line station makes the kampung a shortcut for motorists and pedestrians. The crowds in kampung's neighborhood are thus a combination of residents, tenants, and passers-by.

The kampung's residential area is fairly dense. According to a survey in 2011, the population reached more than 5000 inhabitants. Similar to conditions in Gedong Pompa, one house can be occupied by three families or more because it is also open for rent. The houses are also located close together and separated by narrow alleys with a width of sixty centimeters to two meters. During rush hours, many residents take advantage of the flow of commuting passers-by by selling various items in front of their houses. We also found several open kitchens located on the side of the road.

Activities related to WASH in Cikini Kramat cannot be separated from the existence of Ciliwung River and Kroncong River which are directly adjacent to the residential areas. Due to the limited infrastructure and space, 80.7% of respondents stated that the final disposal of feces is directed to the river. This of course contributes to new problems for the wider water network in the city. However, in contrast to the residents in Gedong Pompa who mostly have to buy water for toilets, residents in Cikini Kramat can still use the well or the pump as the main source of clean water for twenty-nine public toilets. Although some water sources are temporarily or permanently dry during the dry season, residents can still carry over the water using a hose or in buckets to the public toilets.

The availability of water sources from wells or pumps does not necessarily mean that the water can be directly consumed. Limited space affects the distance between the septic tank and the well. Thus, half of the respondents prefer to buy water for consumption due to contamination from the septic tank.

Even though more than half of the residents already use private latrines, public toilets remain an important element because many residents live in rented rooms or rented houses without toilets. This public toilet has been called *mandi, cuci, kakus* (MCK or Bath, Wash, and Toilet) based on their main functions which are different from the public toilet in the public area. MCK has been provided for the residents of specific areas.

In Cikini Kramat, MCKs were built jointly by the residents and have been around since the early days. With the increasing number of kampung residents, more units were added and built together with the help and cooperation of other parties, such as academics in Indonesia and abroad like the academicians from Universitas Indonesia, the University of Tokyo, Chiba University, the University of Florida, Cardiff University, and Chongqing Jiaotong University, The government (LMK-subdistrict forum, PNPM—community empowerment national program 2012, PPMK—urban village empowerment program 2004, etc.) and other communities, like Bank Mandiri. Not only toilets, the construction of RW offices and other public facilities were done through collaboration with similar actors.

The involvement of the other stakeholders does not mean MCKs have no issues. Maintenance and cleanliness are two main issues that happen in MCKs. According to respondents, this is due to the lack of a sense of ownership among the large number of users who use the facility. As a result, many residents blame the residents of the rented rooms and also market vendors who are not local residents. In responding to this problem, it is the women and RT administrators who often take the initiative to clean the toilets. In one unit, residents also implemented a donation box system to help the people who keep the toilets clean.

Similar to Gedong Pompa and other densely populated kampungs, apart from the problem of WASH as a basic human need that must be met, Cikini Kramat also could not escape eviction. Several RTs that used to occupy the area near the Cikini Gold Center have been evicted and moved out a decade ago. Even though infrastructure development such as public toilets, RW offices, and schools in this village have collaborated with academics and other NGOs, this village is still not immune to changes in the spatial plan of Jakarta.

Based on the explanation above, we underline some of the fundamental differences regarding living space and WASH between the two kampungs. One of them is in terms of water sources that can be used for cleaning and consumption. In the context of Gedong Pompa, because of the quality of groundwater that is polluted and contaminated by seawater, the residents get clean water sources for toilets and drinking. 99.3% of respondents in Gedong Pompa village buy water for consumption, and 92% of respondents have to buy water for toilet needs (Pintermidi survey data, 2019). This condition is different from the context of Cikini Kramat village where 98% of respondents used water sourced from the ground (borehole or pump) for toilet purposes without the need to buy it. However, there are also 2% of respondents in Cikini Kramat Village who buy water for toilet purposes and 64% of respondents who buy water for consumption needs. The following is the comparison data of the water problems characteristics, as well as data on the comparison of the distance between the septic tank and the water source in the two kampungs.

From the data in Table 2.1 and Fig. 2.1, the number of respondents who complained about poor water characteristics is low, while the distance between the septic tank and the water source of less than 10 m was observed in almost half of the respondents' answers. However, the researchers emphasize that in terms of WASH, the minimum number of percentages that show poor water characteristics does not indicate a priority or significance scale. As a basic necessity of human life, the percentage of water characteristics that are considered poor must have a closer value to 0 (zero) to ensure all residents can access good water quality without any exception.

Another thing that also marks the difference between the two kampungs is the construction of sanitation facilities or public toilets to accommodate the urgent needs of the residents daily. In the context of Gedong Pompa, the public toilet is built based on the practice of residents' self-reliance. Even though 76.7% of the respondents in the kampung used their latrines, the construction of public toilets is still needed to

Table 2.1 Comparison of water problems characteristic in Cikini Kramat and Gedong Pompa. *Source* PINTERMIDI 2019 survey

Water issues	Cikini Kramat			Gedong Pompa			Both Kampung		
	Yes (%)	No (%)	Total per Kampung (%)	Yes (%)	No (%)	Total per Kampung (%)	Yes (%)	No (%)	Total per Kampung (%)
Turbidity	4.67	95.33	100	4.67	95.33	100	4.70	95.3	100
Color	6.00	94	100	2	98	100	4.00	96	100
Taste	0.70	99.30	100	2	98	100	1.30	98.7	100
Foam	0	100	100	0	100	100	0.00	100	100
Smell	12	88	100	5.30	94.7	100	8.70	91.3	100
None of the above	86	14	100	91.3	8.70	100	88.7	11.3	100

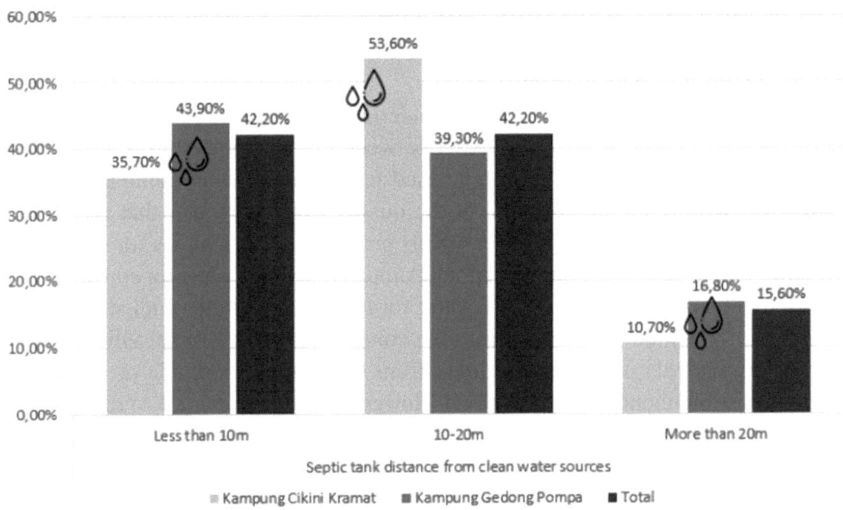

Fig. 2.1 Comparison of septic tank distance from clean water source in Cikini Kramat and Gedong Pompa (*Source* PINTERMIDI 2019 survey)

meet the water and sanitation needs of residents who live in rented rooms and rented houses. On the other hand, in the context of Cikini Kramat, 62.7% of the respondents used private latrines, and 36.7% of the respondents used public or shared latrines. This condition shows that public toilets are a vital infrastructure to meet people's WASH needs. Moreover, public toilets are also used by tenants, traders, and commuters who pass by the kampung on a daily.

2.5.2 Everyday Tactics for Urban Kampung Residents: Social and Cultural Capital for Cikini Kramat and Gedong Pompa Residents

In the midst of various efforts by residents to meet WASH's basic needs, the existence of the two kampung is not immune to the possibility of eviction and relocation due to changes in Jakarta's spatial plan. Since phase 1 of the KIP was implemented in 1969 until the CAP program in recent years, various warnings and relocations have been passed through many crowded kampungs in Jakarta. However, the kampung continues to live and survive. Even though they are considered incompatible with urban modernization, the kampung actually plays an important role in supporting the urban economy by providing affordable housing and various other informal economic activities. This proves that kampung possesses the resilience and potential to support urban liveability and livelihoods.

This resilience and potential are manifested in the various tactics that residents use as a mechanism for their daily survival. The tactic taken is to establish internal–external relations as a socio-cultural capital which is expected to also strengthen the existence of each kampung. Internal tactics are strategies carried out by kampung residents and utilize internal relations with actors in their neighborhoods, while external tactics are strategies that utilize external relations with actors outside their neighborhoods. The tactics that the team observed are the following (Table 2.2):

From the foregoing tactics, we can observe that the two kampungs have attempted to establish relationships with many groups or communities outside their neighborhood. However, the issue of internal relations between residents in the neighborhood cannot escape discussion. Points B-C in Gedong Pompa and points A-B in Cikini Kramat indicate the emergence of tactics that are based on problems in the internal socio-cultural relations of residents, especially related to WASH. To support this indication, the author's team mapped the level of social capital of the residents in the two kampung with the following accumulated results (Fig. 2.2):

Based on the data in the table above, the level of social capital for the residents of the two kampung falls in the low to average levels. This finding is in line with the indication of the problems of internal socio-cultural relations presented in the previous explanation. On the other hand, tactics using external relations do not guarantee success. The tactical use of external relations carried out with other parties outside the environment of the two kampungs (Gedong Pompa: points D and F; Cikini Kramat: points C, D, and E) tends to only answer problems of basic needs that are temporary and does not guarantee certainty of residence nor meet the rights of residents for a place that is suitable for habitation. Only point E in Gedong Pompa and point F in Cikini Kramat are considered to have the potential to provide policy advocacy in ensuring the kampung's geopolitical status. Therefore, the team argues that the internal and external relations carried out by the residents of the two kampungs in general are still less supportive of the ideal long-term urban kampung development. The potential of internal–external networks should be maximized and passed on to

Table 2.2 Internal and external tactics in Cikini Kramat and Gedong Pompa

	Gedong Pompa	Cikini Kramat
Internal tactics	(A) Kampung inhabitants voluntarily and collectively construct public toilets (WASH infrastructure)	(A) The women and neighborhood assoiation's administrators take the initiative to maintain the cleanliness of the public toilets
	(B) Combine the use of water from the well during bathing for the first rinse followed by water from the tank for the final rinse	(B) The use of a donation box for maintaining the cleanliness of public toilets
	(C) Inhabitants who have a private clean water source secure their access to the tap water using locks	
External tactics	(A) Purchasing clean water for use in public toilets and consumption	(A) Kampung inhabitants voluntarily and collectively construct public toilets (WASH infrastructure) collaborating with foreign academics, the government, and NGOs
	(B) Maintain the existence of kampung requires coordinated efforts among multiple levels of local governments, academics,and NGOs such as UPC, JRMK, RUJAK Center of Urban Studies	(B) Purchase clean water for consumption needs
	(C) Maintain the existence of kampung while fulfilling the daily needs of the inhabitants by collaborating with academics in workshops and community work	(C) Construct/revitalize social facilities, such as the community association's office (RW) and early childhood education (PAUD) by collaborating with academics in workshops and community work
		(D) Maintain the existence of kampung by joining LMK (subdistrict council)

the next generation. In other words, the tactics used to address the situation should be more than just solving temporary problems.

2.6 Future Opportunities for Urban Kampung Residents in Jakarta: Opportunities and Hope

There have been many debates over the future of urban kampung, whether it will be eliminated or instead managed as an area with historical values as part of physical evidence of cultural capital that has existed for generations. At the same time, they also have problems that could potentially interfere with their existence in the future. In this case, the team does not intend to create an absolute picture of the future, but

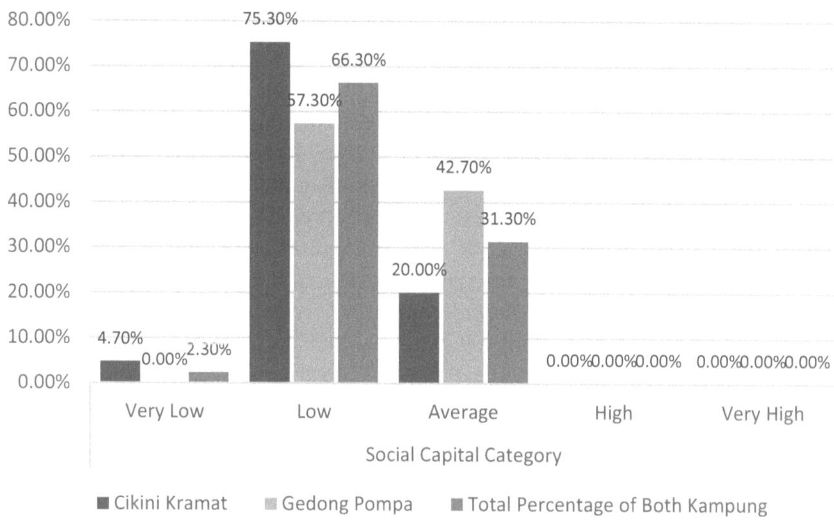

Fig. 2.2 Social capital in Cikini Kramat and Gedong Pompa (*Source* PINTERMIDI 2019 survey)

to the various potentials that an ideal urban kampung can have with ownership rights to space and the fulfillment of basic life needs.

The team also mapped several problems related to space and WASH which were faced by the two kampungs, specifically access to clean water, access to disposal of human waste, and kampung spatial conditions. Regarding access to clean water, the survey results show that only 11.3% of all residents of the two villages still have water issues. This figure is still a problem because clean water is a necessity for everyone. In addition, the condition of clean water requires a large expenditure for kampung residents, especially in the Gedong Pompa area, which is in the coastal area. In the case of human waste disposal, they tend to drain it directly into rivers and into the sea. The space for private activities is also very close to residential areas. This has an impact on their hygiene and health.

Meanwhile, in terms of access to space and settlement, both have their own threats but with significant differences. In Gedong Pompa, which has historically been inhabited by several immigrant families, the historical ties of individuals are slightly weaker than the Cikini Kramat area, which is located in the heart of Jakarta since the colonial period; although, the Cikini Kramat area has also been inhabited by migrants. The cultural nuances of non-Jakarta or non-Betawi newcomers are close to Gedong Pompa, while Cikini Kramat promotes Jakarta and Betawi culture as part of their cultural capital. The historical dimension of the physical artifacts of the two kampungs is also different. In Gedong Pompa, the physical cultural form takes an artificial building, namely the Water Pump Building, which is much younger and has a connection with life history, and its sacred value is different from the Cikini Kramat area which has a tomb and is sacred. Its development, however, is also unclear. Cikini Kramat has a series of cultural activities that reflect the characteristics of the residents inherent in Jakarta and Betawi. Although, activities such as the Independence Day celebrations and carnivals with various regional symbols especially for Betawi people are part of state intervention (Kusno 2020). The characteristics of the city center and periphery, especially on the waterfront, also create different spatial threats. On the waterfront, they have to deal with sea levels that can rise and enable seawater to seep into the area. In contrast to Cikini Kramat, which is in the middle of the city, they are indeed free from the threat of sea level. Even the existence of Cikini Kramat exceeds the tendency of urban villages to move increasingly to the suburbs as studied by Silver (2008). For the threat of eviction, both of them have the same potential. Cikini is in the area of land belonging to the Indonesian State Railway Company (KAI) and Gedong Pompa is in an area that has the potential to experience eviction if there is a change in land use toward industrial or trading interests.

If we map it with a causal-layered analysis, the problems will have implications for the future of urban kampung which looks rim. In terms of political economy, these problems are caused by social, economic, and political inequalities among kampung residents. Many studies also state that the differences between services for urban kampung residents include economic characteristics (Table 2.3).

If we look at a brief mapping of the problems that occurred in the two kampungs, the urban kampung situation seems deadlocked considering the difficulty in building the ideal WASH system as a basic human need, the lack of space, and the possibility to get an eviction. Also, the lack of various types of capital to be exchanged with kampungs development control in urban government at the internal–external level. With this situation, it does not mean that the urban kampung residents cannot survive. They carry out various everyday tactics that reproduce and exchange capital to fulfill their needs.

In Cikini Kramat, residents build networks with various academics and local governments as everyday tactics to increase their social capital. They work together to help with the process of building toilets and other public buildings such as RW offices. This is a manifestation of social capital that is exchanged for resources from related parties, namely academics and other groups. This is also similar to Gedong Pompa which also networks with academics and non-governmental organizations both in assessing and improving space, as well as conducting political contracts with local governments. This can be seen from their daily tactics as given in the previous explanation.

However, the internal and external tactics they carry out daily are only temporary. If it drags on, this condition can affect the kampung's future and even put an end to it. Coupled with the influence of myths about kampung and other parties involved in the tactics, it is believed to prevent kampung from developing in a better direction and continue to trap it in stigma. Kampung residents and other actors need transformative long-term plans. The future transformation process is aimed at achieving a condition where urban kampung can meet the basic needs of WASH and is empowered from an economic-political perspective to avoid eviction and maintain existing capital cultural assets. To carry out the transformation, we need to remap the situation of the two kampungs.

Table 2.3 Causal-layered analysis on the mapping of Cikini Kramat and Gedong. Pompa

Level of analysis	Problematization	Practical implications	Opportunities for alternative solutions
Litany	*Spatial characteristics* – Physical space is increasingly limited, residents build septic tanks close to water sources. – The current land value does not yet provide strong enough incentive for urban kampung's dwellers to prioritize retaining the land over other uses.	– Residents do not have adequate living space. – Health threats – In other words, if neglected, the future will be bleak.	– Spatial planning without losing the right to space and access power. – Sanitation management – Creating a water scheme and equal distribution
	Sanitation and waste management – Management of sanitation that is directly connected to the river and sea waterways which are close to the residential areas. – Compliance to a clean and healthy lifestyle is still low.		

(continued)

Table 2.3 (continued)

Level of analysis	Problematization	Practical implications	Opportunities for alternative solutions
	Access to clean water – Limited water source – Water must be purchased. – Problematic water characteristics		
Systemic	Low distribution of space and services due to a lack of economic, social and cultural capital which are still low. Local government policies that are increasingly less supportive of urban kampung residents, unlike the period during the KIP program.	Residents cannot fulfil their right to space and the need for WASH; they are even threatened to be eliminated.	Reproduce the social and cultural capital of the residents, manage the commons according to the characteristics of each kampung, build collective action with supporting actors. Apart from that, policy advocacy is also needed to strengthen legal-formal efforts.
Worldview	Urban kampung environmental conditions do not meet global health standards. The characteristics of urban kampung are different from the development of global cities.	Urban kampung that does not match the characteristics of a global city will be eliminated. Applying glocalization is only beneficial for kampung that have physical material possession with cultural or passed down historical values	The solution is by creating a new development paradigm and collectivity at the global level
Myth	Myth about kampung	– Urban kampung is a dirty and run-down place. – Urban kampung is a place for poor people with low education. – A place filled with criminals	
	Myths about government	– Sides with marginal groups	
	Myths about NGOs	– Influences and coerces residents	
	Myths about academics and researcher	– Only for research and other academic interests – Temporary	

This tactic of reconstruction becomes more long-term by using back-casting techniques. We set a long-term target for the urban kampung, then we calculate back what is needed. For example, Cikini Kramat and Gedong Pompa want to have access to disposal of human waste with the least possible impact on their health by 2035, so tactical and strategic steps must be made, such as mapping actors who have a positive impact or support programs, identifying emerging issues, and making clear steps. The basic tactical strategy that can be carried out is principally by accumulating social and cultural capital. In contrast to Bourdieu's opinion, which states that the main basis for capital is the economic aspect, kampung residents can optimize the resources of other actors through the exchange of social and cultural capital.

Long-term strategic tactics toward ideal urban kampung transformation are needed to eschew aesthetic solutions, such as physical gentrification, appearance, thematic improvements, or other concepts which do not necessarily suit every kampung. As such, basic needs can be fulfilled on a long-term basis.

2.7 Conclusion

Urban kampungs have existed since colonial times and have become part of a spatial segregation based on class and national ethnicity. During its development, urban kampung became an area for marginalized groups and migrants from various regions. The historical context as well as the recent use of space which places the kampung as a space for the lower-class groups along with the arrangement that is far from the concept of a city has created a debate over the existence of urban kampung. In addition, residents of urban kampung also have to deal with problems of fulfilling their basic needs, such as water, sanitation, and hygiene. As part of the survival mechanism, residents adopt internal and external tactics through various activities which reproduce and exchange capital with various parties. Based on inquiries in Gedong Pompa and Cikini Kramat, the team found that residents are still trapped in a series of temporary tactics. This tactic speaks to the absence of improvisation for the economic and political empowerment of kampung. This of course has implications for the kampung's bleak future and is even threatened to be abolished. Therefore, the team concludes that long-term strategic tactics are needed for the transformation of the urban kampung into part of the ideal urban space. The reconstruction of this tactic is oriented through long-term goals and can be done with the back-casting technique, namely by setting long-term targets for the urban villages and counting down the required requirements. The process of determining the long-term plan requires detailed characteristics and contextual issues in each region. We cannot force the success of one program in a particular village on another. With specific mapping, analysis of strengths, weaknesses, opportunities, and threats, and setting detailed goals, it is possible to carry out a complete long-term transformation. We hope that the kampung residents and stakeholders will set up the plan contextually to move toward a form of urban kampung commoning.

Acknowledgements The researchers would like to thank the PINTERMIDI team, especially the survey team, Gedong Pompa team, and the other Cikini Kramat team members beside us, as well as all residents of Kampung Gedong Pompa and Cikini Kramat who supported us for better quality results. The authors are also grateful for the lesson that has been given by Evawani Ellisa and Raphaella Dewantari Dwianto as our seniors and mentors.

References

Ahn, T. K., and Ostrom, E. 2008. Social capital and collective action. In *The Handbook of Social Capital*, ed., D. Castigliane, J. W. van Deth, and G. Wolleb, 70–100. New York: Oxford University Press.

Aldrich, Daniel P. 2012. Social not physical infrastructure: the critical role of civil society after the 1923 Tokyo. *Earthquake Disasters* 36 (3): 398–419. https://doi.org/10.1111/j.1467-7717.2011.01263.x.

Aldrich, D. P., and M. A. Meyer. 2015. Social capital and community resilience. *American Behavioural Scientist* 59 (2): 254–269. https://doi.org/10.1177/0002764214550299.

Bourdieu, P. 1986. The forms of capital. In *Handbook of theory and research for the sociology of education*, ed. J. Richardson, 241–258. London: Greenwood.

Bunnell, T. 2002. Kampung rules: Landscape and the contested government of urban(e) Malayness. *Urban Studies* 39 (9): 1685–1701. https://doi.org/10.1080/00420980220151727.

Cook, J. 2018. *Imagined futures: Hope, risk and uncertainty*. London: Palgrave Macmillan. https://doi.org/10.1007/978-3-319-65325-9.

Devas, Nick. 1981. Indonesia's Kampung Improvement Program: an evaluative case study. *Ekistics* 48 (286): 19–36. http://www.jstor.org/stable/43620193.

Gottdiener, M. 2012. The new urban sociology. In *Uma ética para quantos*? Vol. XXXIII (Issue 2). https://doi.org/10.1007/s13398-014-0173-7.2.

Inayatullah, S. 2008. Six pillars: Futures thinking for transforming. *Foresight* 10 (1): 4–21. https://doi.org/10.1108/14636680810855991.

Inayatullah, S. 2011. City futures in transformation: Emerging issues and case studies. *Futures* 43 (7): 654–661. https://doi.org/10.1016/j.futures.2011.05.006.

Inayatullah, S. 2013. *Futures studies: Theories and methods*. https://wfsf.org/wp-content/uploads/2020/04/INAYATULLAH-futures_studies_theories_and_methods_published_version_2013_with_pics.pdf

Kawachi, I., Kim, D., Coutts, A., and Subramanian, S. V. 2004. Commentary: reconciling the three accounts of social capital. *International Journal of Epidemiology* 33 (4): 682–690. https://doi.org/10.1093/ije/dyh177

Kusno, A. 2015. Power and time turning: The capital, the state and the kampung in Jakarta. *International Journal of Urban Sciences* 19 (1): 53–63. https://doi.org/10.1080/12265934.2014.992938.

Kusno, A. 2020. Middling urbanism: The megacity and the kampung. *Urban Geography* 41 (7): 954–970. https://doi.org/10.1080/02723638.2019.1688535.

Lumaksono, G. 2013. *Strategi Adaptasi Masyarakat dalam Menghadapi Kekurangan Air Bersih (Studi Kasus di Kampung Jomblang Perbalan Kelurahan Candi Kecamatan Candisari Kota Semarang* (*People's Adaptation Strategies in Facing Clean Water Scarcity, a Case Study of Kampung Jomblang Perbalan, Candi Administrative Village, Candisari Subdistrict*). Unpublished undergraduate thesis. Faculty of Social Sciences, Universitas Negeri Semarang. http://lib.unnes.ac.id/17888/1/3401409002.pdf

Muhtadi, and A. Anggara. 2019. Evaluasi Proses Program Community Action Plan dalam Upaya Meningkatkan Kualitas Lingkungan di Kampung Akuarium Jakarta Utara (Evaluation of Community Action Plan Process in the Effort of Enhancing the Environmental Quality of

Kampung Akuarium, North Jakarta), *Jurnal Al-Ijtimaiyyah, 6*(1), pp. 31–52. https://doi.org/10.22373/al-ijtimaiyyah.v6i1.6674

Nursyahbani, R., and B. Pigawati. 2015. Kajian Karakteristik Kawasan Permukiman Kumuh di Kampung Kota: Studi Kasus Kampung Gandekan Semarang (Analysis of the Characteristics of Slum Housings in Urban Kampung: A Cae Study of Gandekan Kampung, Semarang). *Teknik Perencanaan Wilayah Kota* 4 (2): 267–281. https://doi.org/10.14710/tpwk.2015.8463.

Ostrom, E. 1990. *Governing the commons: The evolution of institutions for collective action.* Cambridge: Cambridge University Press.

Pacione, M. 1990. *Urban problems: An applied urban analysis.* Cambridge: Routledge.

Riogiang, H. 2016. Identifikasi dan Pendampingan untuk Mengatasi Masalah Sanitasi pada Permukiman Kumuh di Kampung Sanger, Sario Manado. *Jurnal LPPM Bidang Sains dan Teknologi, 3*(2), 54–63. https://ejournal.unsrat.ac.id/v3/index.php/lppmsains/article/view/15216/14777

Rolalisasi, A., H. Santosa, and I. Soemarno. 2013. Social capital of urban settlement. *Psychology and Behavioral Sciences* 2 (3): 83–88. https://doi.org/10.11648/j.pbs.20130203.11.

Silver, C. 2008. *Planning the megacity: Jakarta in the twentieth century.* Cambridge: Routledge.

Round, J., C. C. Williams, and P. Rodgers. 2008. Everyday tactics and spaces of power: The role of informal economies in post-Soviet Ukraine. *Social and Cultural Geography* 9365 (May): 171–185. https://doi.org/10.1080/14649360701856110.

Shirleyana, S., S. Hawken, and R. Y. Sunindijo. 2018. City of Kampung: Risk and resilience in the urban communities of Surabaya, Indonesia. *International Journal of Building Pathology and Adaptation* 36 (5): 543–568. https://doi.org/10.1108/IJBPA-02-2018-0025.

Szreter, Simon. 2004. Health by association? Social capital social theory and the political economy of public health. *International Journal of Epidemiology* 33 (4): 650–667. https://doi.org/10.1093/ije/dyh013.

Winayanti, L., and H. C. Lang. 2004. Provision of urban services in an informal settlement: A case study of Kampung Penas Tanggul, Jakarta. *Habitat International* 28 (1): 41–65. https://doi.org/10.1016/S0197-3975(02)00072-3.

Wutich, A. 2011. The moral economy of water Reexamined: Reciprocity, water inscurity, and urban survival in Cochabamba, Bolivia. *Journal of Anthropology Research* 67 (1): 5–26. https://doi.org/10.3998/jar.0521004.0067.102.

Yuwono, S., and S. Wardiningsih. 2016. Mempertahankan Keberadaan Kampung di Tengah Kawasan Modern Jakarta. *Jurnal Arsitektur NALARs* 15 (1): 73–80. https://doi.org/10.24853/nalars.15.1.73-80.

Picture Source:

Freepik. Drops. Bathroom. Retrieved from: https://www.flaticon.com/free-icon/drops_1691455

Chapter 3
The Effects of Household Characteristic Toward Urban Poor's Ability to Pay Domestic Water and Sanitation Services in Urban Kampung Area in Greater Jakarta

Zakianis, Arief Wibisono Lubis, Jenni Anggita, and Rita Damayanti

Abstract This research examines the ability of urban poor in Greater Jakarta to pay for garbage transportation and collection, as well as their willingness to pay for wastewater management. With 717 households involved, it revealed the typical cost of household drinking water and sanitation services was less than or equal to 4% of the minimum wage of the region. Conversely, waste management was still rare (0.3% of the GDP). The total household expenditure per month affected the household's ability to pay for water and sanitation services' fees. The government needs to develop cross-subsidy financing for the urban poor, in terms of water and sanitation budgeting.

Keywords Accessibility · Buying capability · Sanitation

3.1 Introduction

Increased population growth in cities raises various problems, including environmental degradation and urban kampungs resulting from poorly planned city development. It is estimated that Indonesia's proportion of urban dwellers will increase to

Zakianis (✉) · R. Damayanti
Faculty of Public Health, Universitas Indonesia, Depok, Indonesia
e-mail: zakianis@ui.ac.id

A. W. Lubis
Faculty of Economics and Business, Universitas Indonesia, Depok, Indonesia
e-mail: ariefwl@ui.ac.id

J. Anggita
Faculty of Humanities, Jakarta Center for Cultural Studies (JCSS), Universitas Indonesia, Depok, Indonesia

M. Budianta et al. (eds.), *A Liveable Kampung*, Engaging Indonesia,
https://doi.org/10.1007/978-981-96-7866-2_3

56.7% in 2020 and 60% in 2025 (Statistics Indonesia 2013). This increase in urbanization also requires an expansion of urban dwelling areas; however, most urban areas in Indonesia are occupied by unplanned settlements, which are known as urban villages or urban kampungs.

There is no available data on Indonesia's urban kampungs. In Indonesia, according to 2006 data from Statistics Indonesia, 8.3% of households lived in urban kampung areas. In DKI Jakarta Province, 11.02% of households lived in urban kampung areas, while 7.16% of households lived in urban kampung areas in Banten Province. The densely populated urban kampungs promote the spread of various diseases as urban kampungs lack access to adequate sanitation facilities, such as hygienic latrines, and have limited tracts of land, resulting in large quantities of human feces being dumped in open sewers or drainage ditches (Isunju et al. 2011).

Water and sanitation are among the stated Sustainable Development Goals (SDGs) (UNICEF-WHO 2019). Among ASEAN countries, Indonesia ranked second lowest in terms of access to improved water (71%) and lowest in terms of access to improved sanitation (62%) (The ASEAN Secretary 2017). However, these are nationwide figures, and there are usually disparities in access to drinking water and sanitation between urban and rural areas in Indonesia. A total of 81.55% of people in urban areas have access to safe and continuous drinking water sources, while only 64.18% of people in rural areas enjoy such a privilege (Statistics Indonesia 2018). Further, in 48 out of 90 countries, the service coverage for the wealthiest, highest quartile is at least twice as high as that for the poorest, lowest quartile (UNICEF-WHO 2019).

The high rates of open dumping and wastewater disposal into water bodies result in poor sanitation. Only 18.8% of households have a main wastewater disposal system with a closed shelter from the bathroom or washing area, and only 14.3% of households have a main wastewater disposal system with a closed shelter from the kitchen. Waste handling practices are generally still poor as well, with only 34.9% of households getting their waste picked up regularly by professionals (Badan Penelitian dan Pengembangan Kesehatan Kementerian Kesehatan Republik Indonesia 2018).

Public funding can play an important role in helping the government build infrastructure for the sanitation, clean water, and drinking water sectors. However, there is still no information about people's ability to pay the charges for all basic needs in the field of water and sanitation such as waste collection and transportation, clean water, and drinking water, especially in urban areas in Indonesia. There is also no information available on people's willingness to pay (WTP) the charges for household wastewater treatment, especially in households that already have their own fecal management systems. This research aims to measure the ability of poor people living in urban areas, particularly in urban kampungs or *kampongs*, to pay for garbage transportation and collection, as well as the community's WTP wastewater management fees. This information is expected to be useful to the central and regional governments, and other related stakeholders in setting affordable basic tariffs for sanitation, clean water, and drinking water in each area. Additionally, the government can gain a better understanding of urban poor's ability to pay as well as the influence of socio-demographic and political factors on society.

3.2 Economic Factors Affecting Water and Sanitation Services Funding

3.2.1 Economic Loss Due to Water Issues and Lack of Water Funding

Since 2015, the Indonesian government has agreed to allocate IDR 253.8 trillion (USD 18.13 billion) of the national budget for supplying drinking water to urban and rural areas. However, this budget can only fund 50% of the total drinking water supply demand (United Nations-Water Global Analysis and Assessment of Sanitation and Drinking Water [GLAAS] 2019). Due to limited funding, people in rural and urban kampung areas remain profoundly affected by poor access to sanitation (The Economics of Sanitation Initiative [ESI] 2008).

3.2.2 Economic Loss Due to Sanitation Issues and Lack of Sanitation Funding

In Indonesia, the annual financial loss caused by health problems related to poor sanitation is estimated at more than IDR 29 trillion (USD 3.3 billion). Economic loss due to water pollution caused by poor sanitation is estimated at IDR 13 trillion (USD 1.5 billion) annually. Poor sanitation also leads to an annual loss of up to IDR 11 trillion (USD 1.2 billion) in social welfare expenditure due to increased time needed for accessing sanitation, IDR 1.5 trillion (USD 166 million) in the tourism sector, and IDR 0.9 trillion (USD 96 million) in environmental damage due to decreased amounts of fertile land (The Economics of Sanitation Initiative [ESI] 2008).

Economically, regional governments' budget allocations for waste management are still relatively small, with an average of 2.6% of the Regional Government Budget or USD 5–6 spent on waste management per capita per year. This is far below the international standard, which ranges between USD 15 and 20 per capita per year. The budget allocated by the government alone cannot fulfill all communities' needs, considering the limited capacity of the local government (World Bank Group et al. 2018).

In some urban areas worldwide, people spend 0.1–0.7% of GDP per capita on waste management (the United Nations Human Settlements Program 2010). Ideally, waste management expenditure constitutes nearly 0.3–0.5% of GDP. In Dhaka, people spend USD 0.9 per capita per year (0.2% of GDP) on waste management, which is below the set standard. Conversely, in Vienna, the expenditure is USD 137 per capita per year (0.4% of GDP), which is within the standard (International Solid Waste Association n.d.). However, there is no data available on waste management expenditures in Indonesia. Based on a basic specific budget calculation, the waste management expenditure in Indonesia is estimated at IDR 24,000 or USD 1.71

per person per year (Solid Waste Association [InSWA] 2013). From 3 methods of financing available, which are direct–indirect cost, volume-based cost, and weight-based cost (Chung and Yeung 2019), Indonesia adopts a direct financing system (Badan Standardisasi Nasional 1994).

3.3 Conceptual Framework

Previous research related to purchasing ability in the food sector showed demographic factors, such as a household's socio-economic status related to paying ability (Ali and Ali 2020). The ability to pay for water and sanitation is influenced by household characteristics, including the head of the household's gender (Kayaga et al. 2003; Han et al. 2019), an education level (Kayaga et al. 2003; Han et al. 2019), occupation (Kayaga et al. 2003), income (Kayaga et al. 2003; Han et al. 2019; Khan et al. 2014), and house ownership (Kayaga et al. 2003). Education level has a significant relationship with the ability to pay for waste management because education can raise people's environmental awareness and perception on environmental pollution. People with better economic levels have a greater paying ability (Han et al. 2019; Afroz et al. 2009). Higher levels of education and income can be associated with a better understanding of the problem and a greater ability to pay (Han et al. 2019; Afroz et al. 2009).

Based on the theory developed by Kayaga et al. (2003), customer loyalty is influenced by customer satisfaction and other moderating effects. Since sanitation and water are considered as basic needs, including for the urban poor, it is strongly related to the ability to pay for water and sanitation. The conceptual framework developed in this study is modified from Kayaga et al. (2003) (Fig. 3.1).

3.4 Materials and Methods

A cross-sectional study was implemented by involving 717 households living in five urban kampungs of Markisa, Cimone, Nambo, Cikini, and Gedong Pompa in Greater Jakarta. Households participating in this study are chosen through stratified random sampling. The areas have various environmental conditions, including water-catchment peri-urban in the South, bay areas in the North, old urban kampungs in the city center, and the newly industrialized area of Karawaci. All sites have marked issues in terms of waste, water, and health and are subject to prioritized development by the government. The sample size calculation of the estimated proportion was performed with 50% conservative coverage estimation, 8% precision, a design effect of 4, and 95% confidence intervals (Lemeshow 1990). This calculation resulted in 120 samples from each location and 717 observations were considered valid. Since the units of analysis are households, the questionnaire was intended to be answered by the primary caretaker of the household.

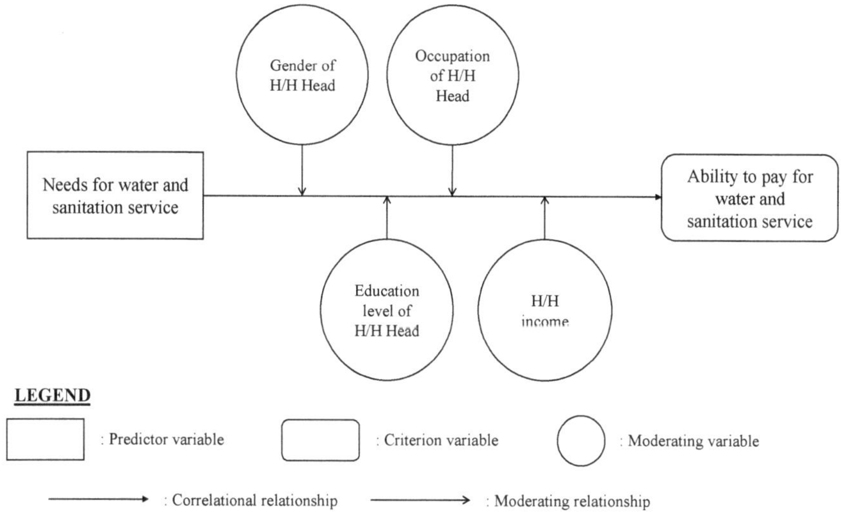

Fig. 3.1 Conceptual framework

3.4.1 Data Collection

Data collected included households' demographic characteristics, waste management practices, drinking water supply, and domestic wastewater treatment. The characteristics of sanitation surveyed included many aspects of waste management, water supply in the household, and liquid waste management in households. These items in the questionnaire were adopted from National Socio-economic Survey and Indonesia Family Life Survey (IFLS). Household expenditure for waste management is calculated based on the costs incurred for transporting waste compared to GDP, which were divided into two groups (poor: $< 0.3\%$ GDP and good: $\geq 0.3\%$ GDP) and water expenses for drinking and toilet use per month, which were also divided into two groups (poor: $> 4\%$ of Provincial Minimum Wage (PMW) and good: $\leq 4\%$ of PMW). A household's WTP for solid waste and stool management was divided into two groups: (1) above the median (> IDR 10,000 or around USD 0.71) and (2) equal to or less than the median ($\leq$ IDR 10,000).

3.4.2 Statistical Analysis

Descriptive statistics are presented in tables. Correlation tests were performed to show correlations, and logistic regression between income and expenditures on water and sanitation was also performed (Hair 2014). To identify factors that can predict people's WTP for waste management, the researchers employed the Ordinary Least Square (OLS) method using a robust standard error to anticipate the heteroscedasticity that can often be found in cross-sectional regressions.

3.5 Results

A significant majority of our respondents were female (nearly 80%). Most of the respondents had finished their high school education, but approximately 12% had not completed elementary school. Kampung Markisa and Cikini were the neighborhoods with the highest proportion of highly educated respondents (50% and 49%, respectively), defined as having graduated from high school or above. Disaggregating the sample according to their geographic location, Gedung Pompa had a significant proportion of respondents without an elementary school degree. However, nearly 5% of the respondents in Cikini possessed at least an undergraduate degree, which was higher than the overall average of 2.9%. Private employees comprised a majority of our sample (almost 32%), but the proportion of those who were self-employed was also high (29.7%). Geographical differences in terms of employment are interesting to see: the proportion of self-employed respondents was significantly lower in Gedung Pompa than in other locations. Meanwhile, Cimone had the highest percentage of private employees (Table 3.1).

Kampung Markisa and Cikini were the neighborhoods with the highest proportion of highly educated respondents (50% and 49%, respectively), defined as having graduated from high school or above. Private employees comprised a majority of our sample (almost 32%), but the proportion of those who were self-employed was also high (29.7%) (Table 3.1).

The average monthly expenditure for each household was IDR 4.86 million (approximately USD 340). Compared to other locations, Nambo recorded the highest average household expenditure at IDR 5.6 million (around USD 401) as well as a standard deviation of IDR 4.0 million (approximately USD 286) (Table 3.2).

The researchers found that most households located in these kampongs did not have a place to collect waste at home, especially those in DKI Jakarta Province (Gedong Pompa and Cikini). Cimone had the worst waste management practices because the proportion of household waste transported by professionals was only 38.1%.

In addition, the researchers found that 38.5% of the water obtained for drinking in the kampungs was branded or refilled bottled water. The sources of water for bathing, washing, and going to the toilet were different from those of water for drinking. Water for daily activities other than drinking was taken from wells. Specifically, in Gedong Pompa, most households (99%) got both drinking water and water for activities by purchasing it. In Markisa, water for drinking could be obtained through the pipeline service, in Gedong Pompa, most households were not served by the piping network and thus could not obtain water in that way. To obtain water, 93.3% of households used non-metered plumbing.

Table 3.1 Household demographic characteristics

Variable	Location					Total (n = 717)
	Markisa	Cimone	Nambo	Cikini	Gedong Pompa	
	n = 148 (%)	n = 147 (%)	n = 123 (%)	n = 149 (%)	n = 150 (%)	
Gender						
Male	44 (29.7%)	18 (12.2%)	19 (15.4%)	31 (20.8%)	37 (24.7%)	149 (20.8%)
Female	104 (70.3%)	129 (87.8%)	104 (84.6%)	118 (79.2%)	113 (75.3%)	568 (79.2%)
Head of household's highest education level						
Low education level	74 (50%)	84 (57%)	74 (60%)	76 (51%)	109 (78%)	417 (58%)
High education level	74 (50%)	63 (43%)	49 (40%)	73 (49%)	41 (27%)	300 (42%)
Occupation						
Fisherman/laborer/scavenger/retired/unemployed	48 (32.4%)	20 (13.6%)	32 (26.0%)	67 (45.0%)	90 (60.0%)	257 (35.8%)
Self-employed	47 (31.4%)	49 (33.3%)	48 (39.0%)	45 (30.2%)	24 (16.0%)	213 (29.7%)
Civil servant/private company worker/public company worker	53 (35.8%)	78 (53.1%)	43 (35.0%)	37 (24.8%)	36 (24.0%)	247 (34.4%)

Most respondents reported that their water was available throughout the year and that most of the water they obtained met their organoleptic requirements by being clear, odorless, and tasteless. For liquid waste management in urban kampungs, most households had their own latrines (84.4%). Among the four kampungs, Gedong Pompa had the highest number of households that did not use latrines, which was 1.3%. The type of toilet used by most households was the gooseneck type equipped with a septic tank (72.7%), which is considered adequate. Most households in Cikini and Gedong Pompa did not have septic tanks and disposed of their waste through sewers instead (Table 3.3).

The average monthly spending for drinking water was approximately IDR 97,000 (USD 6.9), or 1.99% of total monthly spending. The high degree of variation in spending is reflected in the large standard deviation. Households in Gedong Pompa had the highest spending on drinking water, while those in Nambo had the lowest. Reporting this type of spending as a percentage of total spending reveals that residents of Nambo spent a remarkably low proportion of their monthly household expenditures on drinking water (1.17%). In comparison, on average, households in Gedong Pompa spent 2.8% of their monthly expenditures on drinking water.

Table 3.2 Household expenditures per month

Location	Variable			
	Household expenditures			
	Mean	Median	SD	(Min–Max)
Markisa	IDR 4,295,114.92	IDR 3,437,000	IDR 2,994,241.49	(IDR 868,000–IDR 22,209,000)
Cimone	IDR 5,564,058.57	IDR 4,713,000	IDR 3,191,767.79	(IDR 1,225,000–IDR 22,303,000)
Nambo	IDR 5,619,798.44	IDR 4,338,000	IDR 4,044,652.87	(IDR 618,600–IDR 20,914,000)
Cikini	IDR 4,184,728.19	IDR 3,560,000	IDR 2,831,269.11	(IDR 300,000–IDR 22,280,000)
Gedong Pompa	IDR 4,775,650.00	IDR 4,106,250	IDR 2,893,479,77	(IDR 300,000–IDR 18,832,000)
Total ($n = 717$)	IDR 4,860,112.73	IDR 4,040,000	IDR 3,235,423.06	(IDR 300,000–IDR 22,303,000)

The average household spent more on the water for purposes other than drinking than on drinking water (nearly IDR 100,000 or USD 7.14), with a standard deviation that was double the mean. Compared to expenditures on drinking water, there was more variation in expenditures on non-drinking water. Variation associated with the location was common: the average expenditure on non-drinking water in Cikini was only IDR 1544 (approximately USD 0.11), or 0.04% of the average household's total spending, while in Gedong Pompa, this figure reached IDR 368,893 (around USD 26.35) or 7.72% of the average household's total spending.

The average expenditure for waste management was IDR 11,371 (USD 0.81), or 0.23% of the average household's total spending, and the variation in this type of spending was not as high as that in spending for non-drinking water. The highest average spending can be observed in Markisa (IDR 16,453 or USD 1.18), which comprised 0.38% of the average household's total spending. Meanwhile, Cikini had the lowest average spending (IDR 5847 or USD 0.42), which equals 0.14% of the average household's total spending.

For access to toilet facilities for defecation, the average amount which all surveyed households were willing to pay was IDR 22,818 (approximately USD 1.63), with households in Gedong Pompa being willing to pay the highest amount, namely IDR 40,933 (USD 2.92) or 0.86% of their average total monthly spending. However, households in Cikini had the lowest WTP. The average amount that they were willing to pay was IDR 7977 (USD 0.57) or 0.19% of their total monthly spending.

Table 3.3 Sanitation characteristics: waste management, clean water supply, and waste management

Variable	Location					*n* total (717)
	Markisa	Cimone	Nambo	Cikini	Gedong Pompa	
	n = 148 (%)	*n* = 147 (%)	*n* = 123 (%)	*n* = 149 (%)	*n* = 150 (%)	
Waste management						
1. Waste container						
a. Closed container	31 (20.9%)	24 (16.3%)	15 (12.2%)	7 (4.7%)	13 (8.7%)	90 (12.6%)
b. Open container	71 (48.0%)	65 (44.2%)	75 (61.0%)	41 (27.5%)	37 (24.7%)	289 (40.3%)
c. Unavailable	46 (31.1%)	58 (39.5%)	33 (26.8%)	101 (67.8%)	100 (66.7%)	338 (47.1%)
2. Household waste management						
a. Transported by professionals	146 (98.6%)	56 (38.1%)	77 (62.6%)	148 (99.3%)	120 (80%)	547 (76.3%)
b. Not transported by professionals	2 (1.4%)	91 (61.9%)	46 (37.4%)	1 (0.7%)	30 (20%)	170 (23.7%)
3. Frequency of transporting waste by professionals						
a. Daily	114 (77%)	34 (23.1%)	14 (11.4%)	125 (83.9%)	96 (64.0%)	383 (53.4%)
b. Several times a week	30 (20.2%)	21 (14.3%)	49 (39.8%)	17 (11.4%)	23 (15.3%)	140 (19.5%)
c. Once a week	2 (1.4%)	1 (0.7%)	13 (10.6%)	4 (2.7%)	1 (0.7%)	21 (2.9%)
d. Do not know/not transported by professionals	2 (1.4%)	91 (61.9%)	47 (38.2%)	3 (2%)	30 (20%)	173 (24.1%)
4. Waste sorting						
a. Sorted	45 (30.4%)	57 (38.8%)	35 (28.5%)	24 (16.1%)	47 (38%)	218 (30.4%
b. Not sorted	103 (69.6%)	90 (61.2%)	88 (71.5%)	125 (83.9%)	93 (62%)	499 (69.6%)
Drinking water supply						

(continued)

Table 3.3 (continued)

Variable	Location					n total (717)
	Markisa	Cimone	Nambo	Cikini	Gedong Pompa	
	$n = 148$ (%)	$n = 147$ (%)	$n = 123$ (%)	$n = 149$ (%)	$n = 150$ (%)	
5. The main source of drinking water						
a. Bottled water	68 (45.9%)	33 (22.4%)	41 (33.3%)	85 (57.0%)	49 (32.7%)	276 (38.5%)
b. Refilled water	52 (35.1%)	83 (56.5%)	38 (30.9%)	9 (6.0%)	95 (63.3%)	277 (38.6%)
c. Plumbing	11 (7.4%)	5 (3.4%)	8 (6.5%)	5 (3.4%)	6 (4.0%)	35 (4.9%)
d. Drilled wells/pumps/ protected surface waters	17 (11.5%)	26 (17.7%)	33 (26.9%)	50 (33.6%)	0 (0%)	126 (17.6%)
e. Unprotected wells/ surface water	0 (0%)	0 (0%)	3 (2.4%)	0 (0%)	0 (0%)	3 (0.4%)
6. The main way of obtaining drinking water						
a. By purchasing	123 (83.1%)	122 (83.0%)	85 (69.1%)	99 (66.2%)	149 (99.3%)	578 (80.61%)
b. Not by purchasing	25 (16.9%)	25 (17.0%)	38 (30.9%)	50 (33.8%)	1 (0.7%)	139 (19.39%)
7. The main source of water for bathing, washing, and toilets						
Metered plumbing (PAM/PDAM)	45 (30.4%)	29 (19.7%)	33 (26.8%)	1 (0.7%)	4 (2.7%)	112 (16%)
Non-metered plumbing	0 (0%)	0 (0%)	1 (0.8%)	2 (1.3%)	140 (93.3%)	143 (20%)
Drilled wells/pump (protected)/protected spring	103 (69.6%)	113 (76.9%)	81 (65.85%)	146 (98.0%)	6 (4.0%)	441 (62%)

(continued)

Table 3.3 (continued)

Variable	Location					n total (717)
	Markisa	Cimone	Nambo	Cikini	Gedong Pompa	
	n = 148 (%)	n = 147 (%)	n = 123 (%)	n = 149 (%)	n = 150 (%)	
Unprotected wells/ unprotected spring	0 (0%)	5 (3.4%)	8 (6.5%)	0 (0%)	0 (0%)	13 (2%)
8. Main way to obtain water for bathing, washing, and toilets						
a. By purchasing	12 (8.1%)	64 (42.9%)	22 (17.9%)	3 (2.0%)	139 (92.6%)	239 (33.3%)
b. Not by purchasing	136 (91.9%)	84 (57.1%)	101 (82.1%)	146 (98.0%)	11 (7.3%)	478 (66.7%)
9. Water accessibility						
a. Easy	144 (97.3%)	122 (83.0%)	96 (78.0%)	101 (67.8%)	127 (84.7%)	590 (82.3%)
b. Difficult the in dry season	3 (2.0%)	20 (13.6%)	24 (19.5%)	47 (31.5%)	9 (6.0%)	103 (14.4%)
c. Difficult throughout the year	1 (0.7%)	5 (3.4%)	3 (2.4%)	1 (0.7%)	14 (9.3%)	24 (3.3%)
10. Water container availability						
a. Available, open	107 (72.3%)	118 (80.3%)	97 (78.9%)	72 (48.3%)	37 (24.7%)	431 (60.1%)
b. Available, closed	24 (16.2%)	23 (15.6%)	16 (13.0%)	23 (15.4%)	112 (74.7%)	198 (27.6%)
c. Not available	17 (11.5%)	6 (4.1%)	10 (8.1%)	54 (36.2%)	1 (0.7%)	88 (12.3%)
11. Drinking water deficiency incident (minimum 24 h long) in the last 1 month						
a. Yes	3 (2.0%)	2 (1.4%)	2 (1.6%)	3 (2.0%)	20 (13.3%)	30 (4.2%)
b. No/do not know	145 (98.0%)	145 (98.6%)	121 (98.4%)	146 (98%)	130 (86.7%)	687 (95.8%)

(continued)

Table 3.3 (continued)

Variable	Location					*n* total (717)
	Markisa	Cimone	Nambo	Cikini	Gedong Pompa	
	n = 148 (%)	*n* = 147 (%)	*n* = 123 (%)	*n* = 149 (%)	*n* = 150 (%)	
12. Physical condition of water for drinking and cooking in the past 1 month (not turbid, colored, tasteless, foamy, or smelly)						
a. Acceptable	132 (89.2%)	120 (81.6%)	111 (90.2%)	128 (85.9%)	137 (91.3%)	628 (87.6%)
b. Not acceptable	16 (10.8%)	27 (18.4%)	12 (9.8%)	21 (14.1%)	13 (8.7%)	89 (12.4%)
Liquid waste management						
13. Defecation place						
a. Personal latrine	144 (97.3%)	129 (87.8%)	123 (100%)	94 (63.1%)	115 (76.7%)	605 (84.4%)
b. Public latrine	3 (2%)	18 (12.24%)	0 (0.0%)	54 (36.24%)	33 (22%)	108 (15%)
c. No latrine	1 (0.7%)	0 (0.0%)	0 (0.0%)	1 (0.7%)	2 (1.3%)	4 (0.6%)
14. Toilet type for houses with latrine						
a. Gooseneck	146 (98.6%)	146 (99.3%)	121 (98.4%)	148 (99.3%)	116 (77.3%)	677 (94.4%)
b. Non-sanitary latrine	1 (0.7%)	0 (0.0%)	2 (1.6%)	0 (0.0%)	0 (0.0%)	3 (0.4%)
c. Unimproved latrine	0 (0.0%)	1 (0.7%)	0 (0.0%)	0 (0.0%)	32 (21.3%)	33 (4.6%)
15. Fecal final disposal site						
a. Septic tank	144 (97.3%)	144 (98.6%)	98 (79.7%)	28 (18.8%)	107 (71.3%)	521 (72.7%)
b. Wastewater treatment plant	0 (0.0%)	0 (0.0%)	18 (14.6%)	0 (0.0%)	0 (0.0%)	18 (2.5%)

(continued)

Table 3.3 (continued)

Variable	Location					*n* total (717)
	Markisa	Cimone	Nambo	Cikini	Gedong Pompa	
	$n = 148$ (%)	$n = 147$ (%)	$n = 123$ (%)	$n = 149$ (%)	$n = 150$ (%)	
c. Sewers running to pool/field/ river/lake/ sea/others	4 (2.7%)	3 (2.0%)	7 (5.7%)	121 (81.2%)	43 (28.6%)	178 (24.8%)
15. Septic tank distance from water source ($n = 521$)						
a. < 10 m	94 (63.5%)	46 (31.3%)	65 (52.8%)	10 (6.7%)	47 (31.3%)	262 (36.5%)
b. 10–20 m	43 (29.1%)	64 (43.5%)	18 (14.6%)	15 (10.1%)	42 (28.0%)	182 (25.4%)
c. > 20 m	7 (4.7%)	34 (23.1%)	15 (12.2%)	3 (2.0%)	18 (12.0%)	77 (10.7%)
16. Channels for household activities liquid waste disposal						
a. Catchment holes	2 (1.4%)	0 (0.0%)	3 (2.4%)	0 (0.0%)	0 (0.0%)	5 (0.7%)
b. Sewers	140 (94.6%)	141 (95.9%)	76 (61.8%)	2 (1.3%)	112 (74.7%)	471 (65.7%)
c. River/irrigation channel/ lake/sea/ open ground	6 (4.1%)	6 (4.1%)	22 (17.9%)	146 (98.0%)	37 (24.7%)	217 (30.3%)
d. Wastewater treatment plant	0 (0.0%)	0 (0.0%)	22 (17.9%)	1 (0.7%)	1 (0.7%)	24 (3.3%)

The average amount which the respondents were willing to pay for liquid household waste management was IDR 19,386 (USD 1.38), which was less than they were willing to pay for access to toilet facilities for defecation. Similar to the findings regarding defecation facilities, these results show that households in Gedong Pompa had the highest WTP (IDR 34,900 or USD 2.49), while those in Cikini had the lowest (IDR 7852 or USD 0.56).

On average, households were willing to pay IDR 20,140 (USD 1.44) for stool management. The patterns were more or less similar to the previous two findings: households with the highest WTP were those in Gedong Pompa (IDR 30,833 or USD 2.49), and households with the lowest WTP were those in Cikini (IDR 8305 or USD 0.59).

Household expenditures for drinking water and water for sanitary purposes were predominantly acceptable according to the standard, which is less than or equal to 4% of the minimum wage. In most households, however, the waste expenditure was still low at less than 0.3% of GDP (Table 3.4).

The results of this study show that a household's overall monthly expenditure greatly affected the amount that it spent on water for drinking and water for sanitation, as well as the amount that it spent on waste management in the form of waste transportation and garbage collection. The households' WTP for wastewater management was also strongly related to their total monthly household expenditure. Further, the occupation type of the head of household also affected the household's expenditures and WTP (Table 3.5).

3.6 Discussion

The findings suggest that the waste management practices in these urban *kampong*s were relatively good, indicated by the fact that 76.3% of household waste was transported by professionals (Wilson et al. 2015). The percentage of waste transported by professionals in these urban kampungs was lower than that in DKI Jakarta Province (96.5%) (Badan Penelitian dan Pengembangan Kesehatan 2018), where household waste was routinely transported daily by professionals.

We observed that water access was not a crucial problem for most households. Regarding access to safe and continuous drinking water sources, our results are not significantly different from those of the national survey by Statistics Indonesia (2018) (our results show 82%, whereas Statistics Indonesia record shows 81.55%). The quality of toilet facilities was also not a significant problem, as most households used gooseneck toilets and were equipped with septic tanks. Although the water was generally accessible, households that were not equipped with official plumbing overseen by the government tended to find other sources of water on their own. When water is supposed to be delivered by the government but is not, the resulting uncertainty can potentially trigger the development of informal water provider services. This informal water provision has become one of the unofficial systems for city governance in water provision, especially in communities that are not served officially by

Table 3.4 Households' expenditure for drinking water, monthly water usage for household activities, and waste collection and transport fee (GDP/Capita)

Variable	Location					*n* total (717)
	Markisa	Cimone	Nambo	Cikini	Gedong Pompa	
	n = 148 (%)	*n* = 147 (%)	*n* = 123 (%)	*n* = 149 (%)	*n* = 150 (%)	
1. Expenditures for drinking water and other water used for household activities						
a. Bad (> 4% of regional minimum wage)	56 (37.8%)	57 (38.8%)	31 (25.2%)	40 (26.8%)	141 (94.0%)	325 (45.3%)
b. Good (≤ 4% of regional minimum wage)	92 (62.2%)	90 (61.2%)	92 (74.8%)	109 (73.2%)	9 (6.0%)	392 (54.7%)
2. Household waste expenditure per month						
a. Waste expenditure < 0.3% GDP	126 (85.1%)	143 (97.3%)	116 (94.3%)	147 (98.7%)	137 (91.3%)	669 (93.3%)
b. Waste expenditure ≥ 0.3% GDP	22 (14.9%)	4 (2.7%)	7 (5.7%)	2 (1.3%)	13 (8.7%)	48 (6.7%)

Table 3.5 Regression results: predicting households' expenditure for drinking water and waste and willingness to pay for defecation and liquid waste management according to their characteristics

Variable	Water expenditure		Waste expenditure		WTP Defecation		WTP Liquid waste management		WTP Stool management	
	Coefficient	Standard error	Coefficient	Standard error	Coefficient	Standard error	Coefficient	Standard error	Coefficient	Standard error
Number of household members	10.0404***	3.7687	0.1570	0.1644	1.0448	0.5243	0.4540	0.4388	0.5892	0.5603
Children under 5 years old	– 6.1206	13.9510	– 0.5344	0.7024	– 2.8695	2.3990	2.4060	2.2439	3.5885	2.3401
Seniors	– 8.7423	14.6128	0.4667	0.8792	– 3.9497	2.8098	– 0.4664	2.2906	– 2.5685*	1.4920
High education	– 6.9847	12.6061	0.7695	0.6161	– 0.7944	2.5123	1.0803	1.6660	0.6985	2.2648
House ownership	6.9885	12.3363	– 0.0789	0.6246	4.7953	2.2738	1.6874	1.5804	0.3071	1.6089
Income category	21.7157***	5.1357	0.5917**	0.2849	1.8709	1.0081	1.1075	0.8120	2.4978**	1.1818
Non-subsidized electricity	31.5019**	13.4397	2.0471***	0.6293	3.9042	2.4857	1.8518	1.9190	1.8621	1.6841
Markisa	37.9129	15.4704	2.8618***	0.9602	– 1.5709	4.3481	– 3.6857	2.3451	– 3.1600	3.4954
Cimone	14.4865	13.5315	– 4.1587***	1.0007	1.8294	3.5234	0.7191	2.7107	1.9709	4.3898
Cikini	5.3486	14.8842	– 5.935849***	1.2883	– 9.0219***	2.9822	– 8.4910***	2.1392	– 8.1562***	2.9956
Gedong Pompa	346.5258***	22.8502	0.7220	1.0824	17.3199***	5.2085	14.0272***	3.7067	9.0360**	4.3932
Coefficient	– 33.7539	19.2205	7.2170	1.2333	5.3690	3.2837	7.9999	3.2989	6.0280	3.0537
Number of observations	717		717		717		717		717	

(continued)

Table 3.5 (continued)

Variable	Water expenditure		Waste expenditure		WTP Defecation		WTP Liquid waste management		WTP Stool management	
	Coefficient	Standard error	Coefficient	Standard error	Coefficient	Standard error	Coefficient	Standard error	Coefficient	Standard error
R-square	0.4690		0.1702		0.0953		0.1119		0.0941	
Prob *F* statistics	0.0000		0.0000		0.0000		0.0000		0.0000	

*, **, *** represent significant results at 10%, 5%, and 1%, respectively

the government. These kinds of services are mostly permitted by the government; in some cases, they are even required to fulfill the basic water needs of people who are not provided with official public plumbing by the government (Putri et al. 2018).

We found that, overall, an average household's waste-related expenditure was below 0.3% of its GDP, which can be considered a small proportion compared with the international standard (the United Nations Human Settlements Program 2010). Ideally, waste management expenditure constitutes nearly 0.3–0.5% of GDP (International Solid Waste Association 2012). Based on a basic specific budget calculation, the waste management expenditure in Indonesia is estimated at IDR 24,000 or USD 1.71 per person per year (Solid Waste Association (InSWA) 2013). Most households were willing to pay more than IDR 10,000 (approximately USD 0.71) for wastewater treatment. WTP for defecation management was the highest, though it was only slightly higher than that for liquid waste management and stool management.

Some of our most notable findings highlight the associations between differences in WTP and certain demographic characteristics, including income, number of household members, house ownership, and geographical locations. The positive and significant effect of income, except in the case of WTP for waste management, confirms previous findings by Han et al. (2019). A meta-analysis study by Schläpfer and Henley (2006) also found that approximately one-third of studies examining determinants of WTP show the important role of income. This suggests that more affluent households are willing to pay more for these types of expenditures. Meanwhile, the number of household members only affects WTP for defecation management. Indeed, the role of family structure in sanitation-related decision-making has been documented by Dare et al. (2018). House ownership is only significant in the case of WTP for defecation management. One interpretation of this phenomenon is that people's sense of belonging is affected by the defecation system.

The findings on the significance of geographical diversity complement the previous results by Challcharoenwattana and Pharino (2016) and Li et al. (2018) who investigated the importance of geographical factors, which have been overlooked by most other scholars. This disparity might be caused by unequal economic levels among households in these urban kampungs, as suggested by the descriptive data. Consequently, policymakers might consider customizing the amounts charged in different locations.

Meanwhile, there is no significant effect of education level, which is contradictory to the findings by Shahsavar et al. (2019), Danso et al. (2006), and Song et al. (2016), who found that education level is correlated with WTP for waste and environment-related products and/or services.

3.7 Conclusion

The findings show that overall, waste-related expenditure is still under the international standard. Assessments on WTP reveal that most households are willing to pay more than IDR 10,000 (USD 0.71), with the allocation for defecation facilities being the highest. The results also confirm those of previous literature on the significance of several demographic characteristics, especially geographical location and income. Looking at these findings, policymakers might consider customizing water and waste management fees according to locations and income levels.

In addition, central and regional governments should use other data to determine the basic tariffs for sanitation and clean water costs: income, average age, educational background, and employment status (Danso et al. 2006; Afroz et al. 2009; Han et al. 2019). Meanwhile, Jones et al. (2010) confirmed the significance of social capital in influencing people's WTP for solid waste management. Findings related to demographic characteristics have been inconclusive and context-dependent; further studies on this topic in Indonesia can provide interesting insights for policymakers. Any attempt to design an effective policy on waste management must consider these factors. In future studies, it would be interesting to examine households' WTP for different methods of waste treatments in Indonesia, given the limited number of available studies on this issue in contrast to the larger number of previous studies on households' intention to use such services.

Acknowledgements This research was supported by Direktorat Riset dan Pengabdian Kepada Masyarakat Universitas Indonesia (DRPM UI) under PINTERMIDI UI Research Grant No. 945/SK/R/UI/2019. Contribution: We hereby declare that all authors have significantly contributed and agreed with the content of the manuscripts. Conflict of interest: The authors declare that there is no conflict of interest in this study. Availability of data and materials: All data generated or analyzed during this study are included in this published article.

References

Afroz, R., K. Hanaki, and K. Hasegawa-Kurisu. 2009. Willingness to pay for waste management improvement in Dhaka city, Bangladesh. *Journal of Environmental Management* 90 (1): 492–503.

Ali, T., and J. Ali. 2020. Factors affecting the consumers' willingness to pay for health and wellness food products. *Journal of Agriculture and Food Research* 2 (July): 100076. https://doi.org/10.1016/j.jafr.2020.100076.

Badan Penelitian dan Pengembangan Kesehatan, K.K.R.I. 2018. *Hasil utama Riskesdas 2018* (*Main findings of research on basic health*, 2017). Jakarta Indonesia.

Badan Penelitian dan Pengembangan Kesehatan Kementerian Kesehatan Republik Indonesia. 2018. *Laporan nasional Riskesdas* (*National report of research on basic health*).

Badan Standardisasi Nasional. 1994. SNI 19-3964-1994: *Metode pengambilan dan pengukuran contoh timbulan dan komposisi sampah perkotaan.* Badan Standarisasi Nasional.

Challcharoenwattana, A., and C. Pharino. 2016. Wishing to finance a recycling program? Willingness-to-pay study for enhancing municipal solid waste recycling in urban settlements in Thailand. *Habitat International* 51: 23–30.

Chung, W., and I. M. H. Yeung. 2019. Analysis of residents' choice of waste charge methods and willingness to pay amount for solid waste management in Hong Kong. *Waste Management* 96: 136–148. https://doi.org/10.1016/j.wasman.2019.07.020.

Danso, G., P. Drechsel, S. Fialor, and M. Giordano. 2006. Estimating the demand for municipal waste compost via farmers' willingness-to-pay in Ghana. *Waste management* 26 (12): 1400–1409.

Dare, A. M., I. A. Ayinde, A. M. Shittu, D. Akerele, and S. O. Sam-Wobo. 2018. Determinants of the factors affecting willingness to pay for improved sanitation among rural households in Oyo State, Nigeria. *Nigerian Journal of Agricultural Economics* 8 (1): 72–80.

Han, Z., et al. 2019. Public willingness to pay and participate in domestic waste management in rural areas of China. *Resources, Conservation and Recycling* 140 (July 2018): 166–174. https://doi.org/10.1016/j.resconrec.2018.09.018.

International Solid Waste Association. n.d. Solid Waste: guidelines for successful planning, Sao Paolo.

International Solid Waste Association. 2012. *Waste-to-energy: State-of-the-art report* (6th ed.). ISWA.

Isunju, J. B., et al. 2011. Socio-economic aspects of improved sanitation in urban kampungs: A review. *Public Health* 125 (6): 368–376. https://doi.org/10.1016/j.puhe.2011.03.008.

Jones, N., K. Evangelinos, C. P. Halvadakis, T. Iosifides, and C. M. Sophoulis. 2010. Social factors influencing perceptions and willingness to pay for a market-based policy aiming on solid waste management. *Resources, Conservation and Recycling* 54 (9): 533–540.

Jr., J. F. H., et al. (2014). *Multivariate Data Analysis Seventh Ed.* Perason New International Edition.

Kayaga, S., J. Calvert, and K. Sansom. 2003. Paying for water services: Effects of household characteristics. *Utilities Policy* 11 (3): 123–132.

Khan, N. I., R. Brouwer, and H. Yang. 2014. Household's willingness to pay for arsenic safe drinking water in Bangladesh. *Journal of Environmental Management* 143: 151–161. https://doi.org/10.1016/j.jenvman.2014.04.018.

Lemeshow, S., D. W. Hosmer, J. Klar, and S. K. Lwanga. 1990. *Adequacy of sample size in health studies.* John Wiley & Sons.

Li, P., M. H. Chen, Y. Zou, M. Beattie, and L. He. 2018. Factors affecting inn operators' willingness to pay resource protection fees: A case of Erhai Lake in China. *Sustainability* 10 (11): 4049.

Putri, H. T., S. Maryati, and A. N. S. Humaira. 2018. From informal to formal: Status and challenges of informal water infrastructures in Indonesia. *IOP Conference Series: Earth and Environmental Science* 158 (1): 012005. https://doi.org/10.1088/1755-1315/158/1/012005.

Schläpfer, F., and N. Hanley. 2006. Contingent valuation and collective choice, *KYKLOS* 59 (1): 115135.

Shahsavar, T., V. Kubeš, and D. Baran. 2020. Willingness to pay for eco-friendly furniture based on demographic factors. *Journal of Cleaner Production* 250: 119466.

Solid Waste Association (InSWA). 2013. *Menyiasati ongkos pengelolaan sampah. kelopak: sumber informasi kelola sampah dengan bijak* (*Dealing with the cost of waste management, kelopak, information center of wise management of waste*).

Song, Q., Z. Wang, and J. Li. 2016. Exploring residents' attitudes and willingness to pay for solid waste management in Macau. *Environmental Science and Pollution Research* 23: 16456–16462.

Statistics Indonesia. 2013. *Proyeksi penduduk Indonesia 2010–2035* (The projection of Indonesian population 2010–2035). Jakarta.

The ASEAN Secretary. 2017. *ASEAN statistical report on millennium development goals 2017.*

The Economics of Sanitation Initiative (ESI). 2008. *Economic Impacts of A five-country study conducted in Cambodia, Indonesia, Lao PDR, The Philippines, and Vietnam under the Economics of Sanitation Initiative (ESI)*, Jakarta Indonesia.

UNICEF-WHO. 2019. *Progress on household drinking water, sanitation and hygiene I 2000–2017.* New York.

United National-Water Global Analysis and Assessment of Sanitation and Drinking Water (GLAAS). 2019. *The GLAAS 2018/2019 country survey about the GLAAS 2018/2019 country survey WHO policy on the use and sharing of data collected by WHO in member states.*

United Nations Human Settlements Programme. 2010. *Solid waste management in the world's cities, London and Washington DC*.

Wilson, D. C., et al. 2015. "Wasteaware" benchmark indicators for integrated sustainable waste management in cities. *Waste Management* 35: 329–342. https://doi.org/10.1016/j.wasman.2014.10.006.

World Bank Group, Kementerian Koordinator Bidang Maritim & Royal Norwegian Embassy. 2018. *Sampah laut Indonesia* (*Indonesian sea waste*)

Chapter 4
Health Risk Analysis in Urban Kampungs Applying Compost from Organic Waste

Gabriel Andari Kristanto, M. Adhiraga Pratama, Ihsan Ramadhan, and Rusyda Maheswara

Abstract Organic waste, being the largest in terms of composition and generation, is one of the main problems in waste management in the municipal waste stream. Composting can be a proper method to reduce organic waste generation and provide benefits in the form of compost products. However, municipal waste consists of various compositions including hazardous waste, which may contain heavy metals. Therefore, an analysis of health risks in using compost from municipal waste is important. This study aims to analyze the risks of cancer and non-cancer in using organic waste compost sourced from municipal waste in the following urban Kampungs: Cimone and Markisa, Karawaci, Tangerang; Gedong Pompa Penjaringan, North Jakarta; Cikini Kramat, Menteng, Central Jakarta. This research uses a risk assessment method by applying the Monte Carlo analysis on organic waste samples that are contaminated with heavy metals of As, Cd, Hg, and Pb. Kampung Cimone and Markisa each have 0.57 kg/person/day and 0.62 kg/person/day of waste generation, while Gedong Pompa and Cikini Kramat each have 0.68 kg/person/day and 0.74 kg/person/day of waste generation. In Kampong Cimone, Pb of 1.79 mg/kg was detected, and in Markisa, Pb of 2.61 mg/kg and Hg of 5.85 mg/kg were detected, while in Gedong Pompa 1.37 mg/kg of As and 4.49 mg/kg of Pb were detected, and in Cikini Kramat 0.46 mg/kg of Hg was detected. Out of 10,000 cases in Kampong Cimone and Markisa, medium-cancer risks (89.84% and 93.57%) and high non-cancer risks (0.18%, 0.47%, and 99.38%) were obtained. While out of 10,000 cases in Gedong Pompa, the following data were obtained: a high-cancer risk for As (66.68%) and a medium-cancer risk for Pb (59.63%); low non-cancer risks for As (68.78%) and Pb (94.6%). Meanwhile, a high non-cancer risk for Hg (74.39%) was obtained in Cikini Kramat. For the two kampongs in rural settlements, composting can be applied only in Cimone using 16.88% of the generated organic waste, with a potential reduction of waste generation up to 1.54%. For urban settlements, composting can be applied

G. A. Kristanto (✉) · M. A. Pratama · I. Ramadhan · R. Maheswara
Department of Civil Engineering, Faculty of Engineering, Universitas Indonesia, Depok, Indonesia
e-mail: andari@eng.ui.ac.id

M. Budianta et al. (eds.), *A Liveable Kampung*, Engaging Indonesia,
https://doi.org/10.1007/978-981-96-7866-2_4

only in Cikini Kramat using 4.56% of the generated organic waste, with a reduction of waste generation up to 8.96%.

Keywords Organic waste · Urban kampung composting · Health risk · Heavy metal contamination

4.1 Introduction

Daily human activities inevitably generate waste. In Tangerang, a city situated west of Jakarta, these waste generation patterns pose notable challenges; however, they also create valuable opportunities to advance sustainable waste management practices. With a population of 2.2 million people in 2018, Tangerang produces 5687.28 m^3 of waste per day and it is projected to continue to increase (Tangerang Central Bureau of Statistics 2018a, b). In Jakarta, the capital city of Indonesia, waste generation has increased by 30% to 8000 tons per day in 2018, in which the total population of reached over 10 million people in 2017 (Mulyadin et al. 2018; Central Bureau of Statistics 2019). Currently, waste management in the kampungs is only carried out by transporting the waste from sources to landfills. Composting is generally carried out by the informal sector and in very small quantities. However, local governments encourage composting to support agriculture farms in urban areas. In the effort to implement the 3Rs (Reduce, Reuse, Recycle) concept, composting is an effective waste management method to reduce waste generation from the source. In addition, the product obtained from composting can also be reused by the surrounding communities to help fertilize lands. However, it is necessary to pay attention to the level of pollutant concentration in the organic waste to be used as compost material. Research has shown that organic waste mixed with other waste can cause bioaccumulation of heavy metals if the waste is used as a compost base (Pinamonti et al. 1997; Mohee and Soobhany 2014) so the concentration of heavy metals in municipal waste should be considered. The disposal of household hazardous and toxic waste such as cleaning products and their packaging, electronic goods, paint, and used batteries containing lead, cadmium, or mercury without separating the municipal waste can increase the concentration of heavy metals in landfills (Esakku 2003).

When accumulated in the soil, heavy metals can be transferred into the food chain (Page et al. 1987; Masindi and Muendi 2016; Emrotu and Onianwa 2017). It can cause health risks in humans (Wang et al. 2009). These risks are based on the nature of the bioavailability of heavy metals in plants, where there is a process of absorption of heavy metals by plant roots that grow on polluted soil (Ernst 1996). Furthermore, we need to know the toxicity threshold of these heavy metals in long-term exposure, without causing potential effects such as cancer; so that the level of the posed risk can be determined through risk assessments (Dorne et al. 2015). Some heavy metals such as Cr, Cd, As, and Ni are considered to be potentially carcinogenic contaminants, while (Pb), (Hg), and (Mn) are considered non-carcinogenic elements by the International Agency for Research on Cancer (IARC) (IARC 2012). From previous studies, there is strong evidence that arsenic (As) poses cancer risks, and if consumed, it can cause several types of internal cancers such as bladder cancer, kidney cancer, liver cancer, prostate cancer, and lung cancer (Marshall et al. 2007; Morales et al. 2000). In addition to carcinogenic effects, heavy metals may also cause non-carcinogenic effects on health, such as when Cd is consumed into the human body, it affects the blood circulation system which is the main circulatory pathway. If Cd is inhaled, it can cause chronic problems such as impaired lung function and thoracic emphysema. Exposure to Cd can also decrease the sensitivity to smell (Davison et al. 1988; Mascagni et al. 2003).

The nervous system is the most vulnerable to lead poisoning (Tchounwou et al. 2012). Lead exposure to the central nervous system can result in symptoms such as headaches, irritability, memory loss, low attention levels, and dullness (CDC 1991). Exposure to lead should also be given special attention during pregnancy. Lead absorbed by pregnant women can directly flow into the fetus (Ong et al. 1985). In several studies, lead exposure can cause health problems such as respiratory, cardiovascular, kidney, nervous system disorders, and up to the risk of death, which are reviewed at doses of $\pm$ 10 μgPb/dL (ATSDR 2019). In biochemical processes, lead can inhibit or mimic the properties of calcium and interact with proteins. Then lead will bind the minerals and substitute calcium. In addition, lead can also compete with essential metal cations in binding the sites and inhibit the enzyme activity (Flora et al. 2007).

This study aims to analyze cancer and non-cancer health risks associated with the use organic waste compost from municipal solid waste in four *urban* kampungs. The research includes an analysis of waste composition and characteristics, as well as the presence of heavy metals in organic waste. Health risk assessments comprising of cancer and non-cancer risks were conducted using a quantitative approach, employing Monte Carlo simulation techniques to estimate potential exposure outcomes. The findings are intended to inform local government strategies and serve as a reference for enhancing the implementation of the Green Kampung Program.

4.2 Materials and Methods

4.2.1 Overview of Location

This study selected four urban kampungs as research sites to represent diverse urban settlement characteristics. Two of the sites, Gedong Pompa and Cikini Kramat, are located in Jakarta, while the other two, Cimone and Markisa, are situated in Tangerang.

Gedong Pompa is located in Penjaringan, North Jakarta, and encompasses an area of approximately 2.32 hectares. The community hosts around 800 families, with an estimated total population of 1,500 residents. Cikini Kramat, situated in the

Pegangsaan Sub-district of Menteng District, Central Jakarta, covers approximately 1.7 hectares and comprises 13 Neighborhood Units (Rukun Tetangga, RT) (Ellisa 2016). This area is home to an estimated 1,250 families, or approximately 5,000 individuals (Adianto and Iwamoto 2011; Hendrawan 2012; Fijriah and Ellisa 2018). Cimone, a kampung located in Cimone Sub-district, Karawaci District, Tangerang, spans an area of approximately 5 hectares It is inhabited by 88 families with a total population of 252 residents. Meanwhile, Markisa, located in the Pasar Baru Sub-district of the same district, covers 4.1 hectares and is home to 247 families, totaling 1,109 residents.

4.2.2 *Composition and Characteristics of Waste*

The analysis of waste composition and characteristics was conducted by collecting approximately 100 kg of waste samples from Cimone, Markisa, Gedong Pompa, and Cikini Kramat over eight consecutive days. To ensure consistency and reliability in data collection across all study sites, the sampling procedure followed the ASTM D5231–92 standard (ASTM 2016) and SNI 19-3964-1994. The calculation of waste generation was based on the total waste tonnage, truck collection cycles, and the number of residents in each study area. The composition of domestic waste was determined in accordance with the Indonesian National Standard (SNI 19-3964-1994). The classification of household hazardous and toxic waste followed the framework proposed by Tchobanoglous and Kreith (2002), while the categorization of hazardous and toxic waste types was based on the methodology outlined by Delgado et al. (2007). This integrated approach ensured a comprehensive and standardized assessment of waste characteristics across all locations. To assess heavy metal concentrations in organic waste at each study location, researchers randomly collected 1 kg of waste samples daily. The samples were analyzed for arsenic (As), cadmium (Cd), lead (Pb), and mercury (Hg) in accordance with APHA Standard Methods 3112 B and 3120. Mercury levels were measured using the Teledyne Leeman Labs© QuickTrace® M-7600 Mercury Analyzer following the Atomic Absorption Spectroscopy (AAS) method. Arsenic, cadmium, and lead concentrations were determined using the Inductively Coupled Plasma Optical Emission Spectrometry (ICP-OES) method with the Agilent Technologies © Inc Agilent 5100 X instrument. Particularly, cadmium was not detected in samples from any of the four study sites. The use of standardized analytical protocols ensured high accuracy and consistency in the detection of heavy metal concentrations across all locations.

4.2.3 *Calculation of Heavy Metal Risk*

This study uses secondary data from various references. The data from the IFLS (Indonesian Family Life Survey) 2014 in the form of BW (bodyweight) and CR (consumption rate) were adjusted to the characteristics of the community in the four locations. The scenario of exposure to heavy metals is assumed to be in the form of heavy metal contaminants in the compost mixed with the soil, then absorbed (uptaken) by vegetables plants of kales and spinaches, and if these vegetables are consumed by humans, they can pose a risk of cancer and non-cancer. In the risk calculation stage, the concentration of heavy metals in the sample is multiplied by 10,000 random Bv data and the composting ratio, so that 10,000 intake data are obtained. The intake data is multiplied by ED and divided by Tavg and 10,000 random BW data to produce 10,000 dose data. Subsequently, the dose data is multiplied by CSF (Cancer Slope Factor) to calculate cancer risks and RfD (Reference Dose) to produce data on non-cancer risks in the form of 10,000 case frequencies respectively. Next, iterations of mixing of the compost with soil were carried out until a target of 9500 cases (95%) in the low-risk category was obtained. Cancer risk can be determined using the following equation (Dunnivant and Anders 2006):

$$\sqrt{\text{Risk} = ADD \times CSF}$$

The determination of non-carcinogenic risk does not predict the magnitude of the impact that will result from a dose but provides information about an impact that should have occurred. For this reason, the hazard quotient (HQ) is derived from the following equation (Dunnivant and Anders 2006):

$$\sqrt{HQ = \frac{ADD}{RFD}}$$

An HQ value equal to or greater than 1.00 indicates a potential risk of adverse health effect or toxicity. Conversely, an HQ value below 1.00 suggests that the exposure level is unlikely to pose adverse health effects or risks (Fig. 4.1, Tables 4.1, 4.2).

4.3 Results

4.3.1 Generation and Composition of Domestic and Household Hazardous Waste

Waste generation and composition were assessed in four urban kampungs designated as study sites: Cimone, Markisa, Gedong Pompa, and Cikini Kramat (Rosesar and Kristanto, 2020). The average daily waste generation rates were 0.57 kg/person in Cimone, 0.62 kg/person in Markisa, 0.68 kg/person in Gedong Pompa, and 0.74 kg/person in Cikini Kramat. The domestic waste composition, classified into paper/cardboard, plastic, textiles, absorbent hygiene products, wood, rubber, styrofoam, glass, organic matter, metal, e-waste, and hazardous and toxic waste, consistently

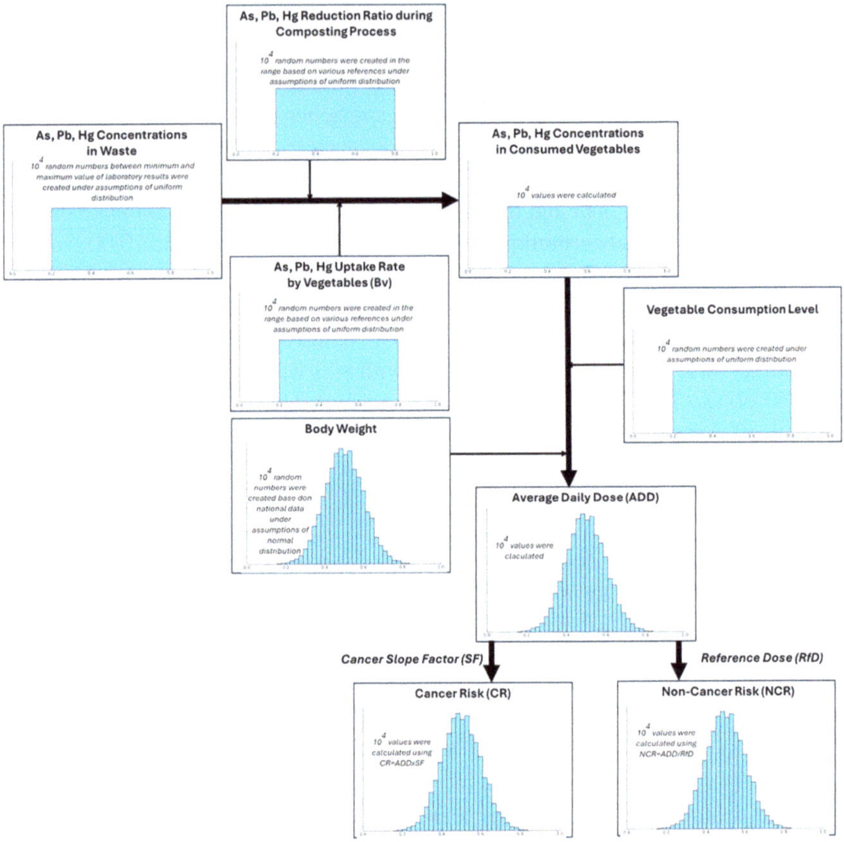

Fig. 4.1 Flowchart of the Monte Carlo Simulation Scheme for evaluating chemical exposure and associated health risks (Mooney, 1997).

showed organic waste as the dominant fraction. The respective average proportions of organic waste were 70.30% in Cimone, 60.45% in Markisa, 56.53% in Gedong Pompa, and 59.21% in Cikini Kramat. Household hazardous and toxic waste generation rates remained relatively low, with 0.003 kg/person/day in Cimone, 0.001 kg/person/day in Markisa, 0.005 kg/person/day in Gedong Pompa, and 0.001 kg/person/day in Cikini Kramat. The dominant hazardous and toxic waste types across all locations included packaging from household cleaning products (42.14% in Cimone, 41.8% in Markisa, 30.62% in Gedong Pompa, and 29.94% in Cikini Kramat), personal care product packaging (34.08%, 20.40%, 32.53%, and 47.70%, respectively), and pharmaceutical packaging (11.94%, 22.32%, 11.51%, and 16.47%, respectively). (Table 4.3). These findings emphasize the critical role of waste segregation and targeted strategies for both organic and hazardous and toxic waste which eventually contribute to improved environmental health and sustainability in densely populated urban kampungs.

Table 4.1 Secondary data from Jakarta Province specifically focusing on Kampung Gedong Pompa and Kampung Cikini Kramat.

No.	Parameter	Mean	Median	Min	Max	References
1	Consumption rate (CR)					IFLS (2014a, b)
	Water spinach (kg)	0.66	0.66	0	1.33	
	Spinach (kg)	0.29	0.29	0	0.58	
2	Duration of exposure (ED)					Fjeld et al. (2007)
3	Body weight (BW) (kg)	59.77	58	30.5	99.7	IFLS (2014a, b)
4	Fraction of digested food (FI)	100%	100%	100%	100%	US. EPA (2000)
5	Average exposure time (AT) (year)	70	70	70	70	Fjeld et al. (2007)
6	*Cancer slope factor* (CSF)					Wang et al. (2017) and OEHHA (2007)
	As (mg/kg day)$^{-1}$	1.5	1.5	1.5	1.5	
	Pb (mg/kg day)$^{-1}$	0.0085	0.0085	0.0085	0.0085	
	Hg (mg/kg day)$^{-1}$	N/A	N/A	N/A	N/A	
7	*Reference Dose* (RfD)					ATSDR (2009, 2010, 2018, 2019)
	As (mg/kg day)	0.003	0.003	0.003	0.003	
	Pb (mg/kg day)	0.0035	0.0035	0.0035	0.0035	
	Hg (mg/kg day)	0.0004	0.0004	0.0004	0.0004	
8	Bioaccumulation factor of heavy metals (Bv)					Fjeld et al. (2007) and Zhou et al. (2016)
	As	0.0105	0.0105	0.01	0.011	
	Pb	0.0086	0.0087	0.00076	0.0165	
	Hg	0.1638	0.1628	0.03	0.3	
9	Composting effect ratio					Whittle and Dyson (2002), Isrun et al. (2018) and Smolinska (2015)
	As	0.79	0.79	0.09	1.5	
	Pb	0.67	0.67	0.07	1.29	
	Hg	0.78	0.78	0.28	1.3	

Table 4.2 Secondary data from Banten Province specifically focusing on Kampung Cimone and Kampung Markisa

No.	Parameter	Mean	Median	Min	Max	References
1	Consumption rate (CR)					IFLS (2014a, b)
	Water spinach (kg)	0.288	0.289	0	0.579	
	Spinach (kg)	0.139	0.137	0	0.279	
2	Duration of exposure (ED)	65	65	65	65	Fjeld et al. (2007)
3	Body weight (BW) (kg)	43.47	43.29	2.52	84.65	IFLS (2014a, b)
4	Fraction of digested food (FI)	100%	100%	100%	100%	US. EPA (2000)
5	Average exposure time (AT) (year)	70	70	70	70	Fjeld et al. (2007)
6	*Cancer slope factor* (CSF)					Wang et al. (2017) and OEHHA (2007)
	As $(\text{mg/kg day})^{-1}$	1.5	1.5	1.5	1.5	
	Pb $(\text{mg/kg day})^{-1}$	0.0085	0.0085	0.0085	0.0085	
	Hg $(\text{mg/kg day})^{-1}$	N/A	N/A	N/A	N/A	
7	*Reference Dose* (RfD)					ATSDR (2009, 2010, 2018, 2019)
	As (mg/kg day)	0.003	0.003	0.003	0.003	
	Pb (mg/kg day)	0.0035	0.0035	0.0035	0.0035	
	Hg (mg/kg day)	0.0004	0.0004	0.0004	0.0004	
8	Bioaccumulation factor of heavy metals (Bv)					Fjeld et al. (2007) and Zhou et al. (2016)
	As	0.00835	0.00836	0.00076	0.016	
	Pb	0.1635	0.1631	0.03	0.3	
9	Hg					Whittle and Dyson (2002), Israun et al. (2018) and Smolinska (2015)
	Composting effect ratio	0.67	0.67	0.07	1.29	
	As	0.78	0.78	0.28	1.3	

Table 4.3 Concentration of heavy metals in organic waste

No. sample	Location	Heavy metal (mg/kg organic waste)			
		As	Cd	Pb	Hg
1	Cimone	< 1.00	< 1.00	< 1.00	< 0.05
2		< 1.00	< 1.00	< 1.00	< 0.05
3		< 1.00	< 1.00	1.79	< 0.05
4		< 1.00	< 1.00	< 1.00	< 0.05
5		< 1.00	< 1.00	< 1.00	< 0.05
1	Markisa	< 1.00	< 1.00	< 1.00	< 0.05
2		< 1.00	< 1.00	< 1.00	< 0.05
3		< 1.00	< 1.00	< 1.00	< 0.05
4		< 1.00	< 1.00	2.61	< 0.05
5		< 1.00	< 1.00	< 1.00	5.85
1	Gedong Pompa	< 1.00	< 1.00	< 1.00	< 0.05
2		1.37	< 1.00	< 1.00	< 0.05
3		< 1.00	< 1.00	< 1.00	< 0.05
4		< 1.00	< 1.00	< 1.00	< 0.05
5		< 1.00	< 1.00	4.49	< 0.05
1	Cikini Kramat	< 1.00	< 1.00	< 1.00	< 0.05
2		< 1.00	< 1.00	< 1.00	< 0.05
3		< 1.00	< 1.00	< 1.00	< 0.05
4		< 1.00	< 1.00	< 1.00	< 0.05
5		< 1.00	< 1.00	< 1.00	0.46

4.3.2 Heavy Metals Concentration Test Results

4.3.3 Cancer Risk

In Cimone, the distribution of cancer risk associated with lead (Pb) exposure was successfully determined, revealing that the majority of cases for both males and females fall within the medium-risk category (Tables 4.4 and 4.5). Males exhibited a slightly higher proportion of medium-level cancer risk compared to females. Among 10,000 simulated cases, a cancer risk value of 10^{-6} was observed in 5,346 male and 5,523 female cases, corresponding to probabilities of 53.46% and 55.23%, respectively—indicating a 1 in 1,000,000 chance of developing cancer. A cancer risk value of 10^{-5} was recorded in 3,425 male and 3,286 female cases, translating to probabilities of 34.25% and 32.86%, or a 1 in 100,000 chance of cancer occurrence. Additionally, a cancer risk value of 10^{-4}, representing a 1 in 10,000 chance, was identified in 213 male and 84 female cases, corresponding to probabilities of 2.13% and 0.84%, respectively. Lower cancer risk levels were also identified. A risk value of 10^{-7} was observed in 982 male and 1,074 female cases, indicating probabilities of 9.82% and 10.74% (1 in 10,000,000 chance), while a risk value of 10^{-8} was found in 34 male and 33 female cases, reflecting probabilities of 0.34% and 0.33% (1 in 100,000,000 chance). In particular, no cases in either sex were categorized within the high-risk group, indicating a 0.00% probability of high cancer risk.

In Markisa, the distribution of cancer risk associated with lead (Pb) exposure was successfully assessed, revealing that the majority of cases in both males and females fall within the medium-risk category. Males exhibited a slightly higher proportion of medium-level cancer risk compared to females. Among 10,000 simulated cases, a cancer risk value of 10^{-6} was observed in 4,611 male and 4,800 female cases, indicating probabilities of 46.11% and 48.00%, respectively—corresponding to a 1 in 1,000,000 chance of developing cancer. A cancer risk value of 10^{-5} was recorded in 4,334 male and 4,296 female cases, reflecting probabilities of 43.34% and 42.96%, respectively, or a 1 in 100,000 cancer risk. In addition, a cancer risk value of 10^{-4}—equivalent to a 1 in 10,000 probability—was found in 412 male and 182 female cases, corresponding to 4.12% and 1.82% of the total. Lower cancer risk levels were also present. A cancer risk value of 10^{-7} appeared in 632 male and 707 female cases, indicating probabilities of 6.32% and 7.07%, respectively (1 in 10,000,000). Meanwhile, a cancer risk value of 10^{-8} was observed in 11 male and 15 female cases, representing probabilities of 0.11% and 0.15%, respectively (1 in 100,000,000). Importantly, no cases in either sex were categorized in the high-risk group, indicating a 0.00% probability of high cancer risk. The cancer risk associated with mercury (Hg) exposure could not be determined, as the Cancer Slope Factor (CSF) for (Hg) has not yet been established. Current research classifies (Hg) as non-carcinogenic (U.S. EPA 1995).

In Gedong Pompa, the research evaluated the distribution of cancer risk associated with arsenic (As) exposure. The findings indicate that the majority of cancer risk cases in both males and females fall within the high-risk category (10^{-3}–10^{-2}), with females exhibiting a slightly higher proportion of high-level cancer risk than males. Among 10,000 simulated cases, a cancer risk value of 10^{-3} was observed in 6,922 male and 6,971 female cases, corresponding to probabilities of 69.22% and 69.71%, respectively—reflecting a 1 in 1,000 chance of developing cancer. Additionally, a cancer risk value of 10^{-2}, representing a 1 in 100 probability, was recorded in 1,047 male and 1,174 female cases, yielding probabilities of 10.47% and 11.74%, respectively. A smaller proportion of cases fell within the medium-risk category. A cancer risk value of 10^{-4} was identified in 1,990 male and 1,820 female cases (19.90% and 18.20%), while a risk value of 10^{-5} occurred in 41 male and 35 female cases (0.41% and 0.35%). No cases were classified within the low-risk group for either sex, indicating a 0.00% probability of low cancer risk. Regarding lead (Pb) exposure, the distribution analysis revealed that most cancer risk cases in both males and females fall within the medium-risk category. Females demonstrated a higher proportion of medium-level cancer risk compared to males, highlighting a sex-based difference in exposure outcomes.

Among 10,000 simulated cases, the distribution of cancer risk values for both males and females reflects a predominance in the moderate-risk category. A cancer

Table 4.4 Distribution of cancer risk for men

Risk level	Risk of cancer	Cimone (Pb)		Markisa (Pb)		Gedong Pompa (As)		Gedong Pompa (Pb)	
		Frequency	Probability (%)	Frequency	Probability (%)	Frequency	Probability (%)	Frequency	Probability (%)
Low risk	10^{-8}	34	0.34	11	0.11	0	0	1	0.01
	10^{-7}	982	9.82	632	6.32	0	0	120	1.2
Medium risk	10^{-6}	5346	53.46	4611	46.11	0	0	2266	22.66
	10^{-5}	3425	34.25	4334	43.34	41	0.41	6016	60.16
	10^{-4}	213	2.13	4.12	4.12	1990	19.9	1584	15.84
High risk	10^{-3}	0	0.00	0	0.00	6922	69.22	13	0.13
	10^{-2}	0	0.00	0	0.00	1047	10.47	0	0

Table 4.5 Distribution of cancer risk for women

Risk level	Risk of cancer	Cimone (Pb)		Markisa (Pb)		Gedong Pompa (As)		Gedong Pompa (Pb)	
		Frequency	Probability (%)	Frequency	Probability (%)	Frequency	Probability (%)	Frequency	Probability (%)
Low risk	10^{-8}	33	0.33	15	0.15	0	0	0	0
	10^{-7}	1074	10.74	707	7.07	0	0	111	1.11
Medium risk	10^{-6}	5523	55.23	4800	48.00	0	0	2119	21.19
	10^{-5}	3286	32.86	4296	42.96	35	0.35	6089	60.89
	10^{-4}	84	0.84	182	1.82	1820	18.2	1667	16.67
High risk	10^{-3}	0	0.00	0	0.00	6971	69.71	14	0.14
	10^{-2}	0	0.00	0	0.00	1174	11.74	0	0

risk value of 10^{-6} was observed in 2,266 male and 2,119 female cases, corresponding to probabilities of 22.66% and 21.19%, respectively—indicating a 1 in 1,000,000 chance of developing cancer. The most frequently observed risk value was 10^{-5}, recorded in 6,016 male and 6,089 female cases, with probabilities of 60.16% and 60.89%, respectively, representing a 1 in 100,000 cancer risk. In addition, a cancer risk value of 10^{-4}—equivalent to a 1 in 10,000 probability—was found in 1,584 male and 1,667 female cases, accounting for 15.84% and 16.67% of the total, respectively. Other risk levels fell within both the low- and high-risk categories. A cancer risk value of 10^{-8} was identified in 1 male and 120 female cases, indicating probabilities of 0.01% and 1.20%, respectively (1 in 100,000,000 chance). A risk value of 10^{-7} was observed in 120 male and 111 female cases, corresponding to probabilities of 1.20% and 1.11% (1 in 10,000,000 chance). At the higher end of the risk spectrum, a cancer risk value of 10^{-3}—representing a 1 in 1,000 probability—was detected in 13 male and 14 female cases, with associated probabilities of 0.13% and 0.14%, respectively. These results provide a comprehensive understanding of cancer risk distribution across risk levels and genders.

4.3.4 Non-cancer Risk

Based on Table 4.4, the analysis of rural settlements indicates that the frequency of "Hazard Quotient" (HQ) values for both men and women in the selected kampungs remained below 1.00, signifying a low non-cancer risk, generally means that the likelihood of experiencing harmful health effects other than cancer due to exposure to a substance or situation is very low. This is determined by comparing the level of exposure to a reference level using HQ. If the HQ is less than 1.00, it suggests that non-cancer health effects are unlikely. This outcome exceeds the 95% probability threshold, with more than 9,500 out of 10,000 cases meeting the low-risk criterion. In Cimone, the distribution of non-cancer risk due to lead (Pb) exposure further supports this favorable outcome. Among male cases, 9,982 out of 10,000 (99.82%) were categorized as having a low non-cancer risk, while only 18 cases (0.18%) exhibited a high non-cancer risk. For female cases, the results were even more encouraging: all 10,000 cases (100.00%) were classified as low non-cancer risk, with zero instances of high-risk outcomes. These findings underscore the generally safe exposure levels in rural settings, particularly for women, with minimal indications of elevated health risk from lead.

The distribution of non-cancer risk due to lead (Pb) exposure in Markisa demonstrated predominantly low-risk outcomes. Among 10,000 simulated cases for males, 9,953 cases (99.53%) were categorized as having a low non-cancer risk, while only 47 cases (0.47%) exhibited a high non-cancer risk. Similarly, for females, 9,998 cases (99.98%) were classified as low risk, with just 2 cases (0.02%) falling into the high-risk category. Although the average dose of (Pb) exposure was slightly higher for males than for females, the exposure levels for both sexes remained below the established reference dose (RfD), indicating a generally safe exposure profile. In contrast, the distribution of non-cancer risk associated with mercury (Hg) exposure in Markisa revealed a higher proportion of cases exceeding the RfD threshold. For males, 9,938 cases (99.38%) out of 10,000 showed a high non-cancer risk (HQ > 1.00), while only 62 cases (0.62%) were classified as low risk. Similarly, for females, 9,935 cases (99.35%) were categorized as high non-cancer risk, with just 65 cases (0.65%) falling into the low-risk category. These findings suggest that mercury exposure levels in both sexes exceed the RfD in the vast majority of cases, highlighting a potential area of concern for public health monitoring and intervention.

In urban settlements, the distribution of Hazard Quotient (HQ) values due to arsenic (As) exposure in Gedong Pompa revealed that the majority of cases for both males and females were categorized as having a low non-cancer risk (HQ less than 1.00). However, the frequency of low-risk cases was below the 95% probability threshold (i.e., 9,500 out of 10,000 cases). Specifically, 7,089 male cases (70.89%) and 6,910 female cases (69.10%) out of 10,000 fell into the low-risk category, while 2,911 (29.11%) and 3,090 (30.90%) cases, respectively, were classified as high non-cancer risk. In contrast, the non-cancer risk due to lead (Pb) exposure presented a more favorable outcome. For males, 9,729 cases (97.29%) demonstrated low non-cancer risk, with only 280 cases (2.80%) showing high non-cancer risk. Similarly, for females, 9,624 cases (96.24%) were in the low-risk category, while 376 cases (3.76%) were identified as high risk. These results suggest that (Pb) exposure in Gedong Pompa remains largely within safe limits for the majority of the population, while As exposure warrants further attention to reduce elevated health risks.

The distribution of non-cancer risk associated with mercury (Hg) exposure in Cikini Kramat revealed that the majority of cases for both males and females exceeded the reference threshold (HQ $\geq$ 1.00), indicating a high non-cancer risk. The frequency of HQ values below 1.00, which reflects low non-cancer risk, was lower than the 95% probability target (i.e., fewer than 9,500 out of 10,000 cases). Specifically, 2,602 male cases (26.02%) were identified as having a low non-cancer risk, while 7,398 cases (73.98%) were classified as high non-cancer risk. Among females, 2,391 cases (23.91%) exhibited a low non-cancer risk, compared to 7,609 cases (76.09%) with a high non-cancer risk, as shown in Table 4.6.

4.4 Discussion

4.4.1 Generation and Composition of Domestic and Household Hazardous and Toxic Waste

The waste generation rate in Cimone is lower than the average rate reported for Tangerang, which stands at 0.58 kg/person/day. In contrast, Markisa exhibits a higher waste generation rate. These differences can be attributed to various influencing factors, including demographic characteristics, economic conditions, and patterns of consumer behavior (Masebinu et al. 2017; Bello 2018; Liu et al. 2019; Monavari

Table 4.6 Distribution of non-cancer risk frequency for arsenic (As), lead (Pb), and mercury (Hg).

Location	Heavy metal	Gender	Frequency	
			HQ < 1.00	HQ > 1.00
Kampung Cimone	Pb	Male	9982	18
		Female	10,000	0
Kampung Markisa	Pb	Male	9953	47
		Female	9998	2
	Hg	Male	62	9938
	Hg	Female	65	9935
Kampung Gedong Pompa	As	Male	7089	2911
		Female	6910	3090
	Pb	Male	9720	280
		Female	9624	376
Kampung Cikini Kramat	Hg	Male	2602	7398
		Female	2391	7609

et al. 2012). In addition, Cimone would be expected to generate more waste than Markisa given its greater community access to food and beverage vendors. However, the observed differences may be influenced by additional factors not addressed in this study. Cimone recorded a higher proportion of organic waste (70.30%) compared to Markisa (60.45%). The primary contributor to organic waste is food waste, which is strongly associated with residents' accessibility to food stalls and informal eateries.

The household hazardous and toxic waste generation rates in Cimone and the comparison site were evaluated to contextualize local waste management challenges. The analysis revealed that the generation rate of household hazardous and toxic waste in Cimone exceeds the average rate reported in Sleman Regency, Central Java, Indonesia, which stands at 0.002 kg/person/day (Iswanto et al. 2016). In contrast, Markisa exhibits a lower household hazardous and toxic waste generation rate compared to the average reported in Sleman Regency, Central Java. Moreover, the household hazardous and toxic waste generation rates in both Cimone and Markisa remain lower than those observed in several developed countries, including Japan and Germany, where the average generation rate reaches approximately 0.005 kg/person/day (Yasuda and Tanaka 2006; Nelles et al. 2016), and Denmark with an average generation rate of 0.014 kg/person/day (Fjelsted and Christensen 2007). The magnitude of each of these generation rates can be influenced by several factors, such as the sampling methodology applied in the study, the classification of household hazardous and toxic waste, and the waste management plan (Ojeda-Benitez et al. 2013). The composition of household hazardous and toxic waste in the study areas revealed a significant proportion of packaging from personal care and household cleaning products. When compared with findings from two regions in Mexico, as reported in the study by Delgado et al., similar patterns in hazardous and toxic waste composition were observed (Delgado et al. 2007). The significant presence of personal care and household cleaning product packaging in both kampungs can be attributed to the community's limited financial resources. As a result, residents tend to purchase affordable products in low-cost sachet packaging.

A comparative analysis between Gedong Pompa and Cikini Kramat revealed that Cikini Kramat generates a higher volume of waste, with an average generation rate of 0.746 kg/person/day, compared to 0.613 kg/person/day in Gedong Pompa. In both areas, organic waste constituted the largest proportion of total waste, accounting for 59.21% in Cikini Kramat and 56.53% in Gedong Pompa. The higher percentage of organic waste in Cikini Kramat is likely influenced by its larger population size. Cikini Kramat comprises approximately 1,250 households, whereas Gedong Pompa consists of around 800 households.

4.4.2 Heavy Metal Concentration Test Results

Lead in organic waste sample no-3 in Cimone and organic waste sample no-4 in Markisa is associated with personal cleaning products like soaps in general, such as cosmetic and bath soaps (Alizadeh et al. 2017; Endah and Surantaatmadja 2019). Mercury in organic waste no-5 is also associated with personal cleaning products and cleaning products such as conventional powdered detergents (Alizadeh et al. 2017; Allen 1995). Arsenic in organic waste sample no-2 in Kampung Gedong Pompa is associated with personal cleaning products such as cosmetic ingredients (Saidalavi et al. 2017), while lead in organic waste sample no-5 is associated with electronic waste where lead is often found in industrial manufacturing processes (Korzun and Heck 1990; Herat 2008). Mercury in organic waste sample no-5 is associated with electronic waste, where mercury is often used in LCD panels and electronic components such as fluorescent tubes, circuits, and switches (Randall 1995; Chen et al. 2011).

4.5 Risk Characterization

4.5.1 Cancer Risk

The target threshold for low cancer risk, defined as a probability of less than 10^{-6}, was set at 95% of the population in the risk calculation. When this target was not achieved, iterative dilution was conducted by blending organic compost containing heavy metals with compost presumed to be free of such contaminants. In Cimone, lead (Pb) exposure resulted in a low cancer risk in 6,362 out of 10,000 cases (63.62%). Iteration results yielded a (Pb) concentration of 0.30 mg/kg in the organic compost, representing 16.88% of the initial (Pb) concentration. Similarly, in Markisa, (Pb) exposure led to a low cancer risk in 5,480 cases (54.80%), with the corresponding

(Pb) concentration in the compost measured at 0.30 mg/kg, equivalent to 11.58% of the initial concentration. In Gedong Pompa, simulations of arsenic (As) exposure revealed a high cancer risk in 6,668 cases (66.68%). The (As) concentration in the compost, based on iteration outcomes, was 0.054 mg/kg—just 0.0395% of the initial (As) concentration. For (Pb) exposure in the same area, a medium cancer risk was observed in 5,963 cases (59.63%), and the corresponding (Pb) concentration in compost was 0.141 mg/kg, which is 3.15% of the initial (Pb) content. In contrast, cancer risk estimation for mercury (Hg) in Cikini Kramat could not be performed due to the absence of a Cancer Slope Factor (CsF) for (Hg).

4.5.2 *Non-cancer Risk*

A low non-cancer risk due to lead (Pb) contamination in 94.6% of cases, with an estimated (Pb) concentration of 4.2 mg/kg based on iteration results, representing 93.5% of the initial (Pb) concentration. In Cikini Kramat, mercury (Hg) exposure resulted in a high non-cancer risk in 25.61% of cases. The corresponding (Hg) concentration obtained from the iteration was 0.02 mg/kg, accounting for 4.56% of the initial (Hg) concentration.

4.5.3 *Technical Recommendations*

Implementing a waste management hierarchy serves as an effective strategy for reducing overall waste volume. Segregating waste at the source is essential to prevent contamination between waste components, thereby minimizing potential health risks and environmental pollution (Singh et al. 2018; Ferrari et al. 2016). Following the characterization of both cancer and non-cancer risks associated with heavy metal exposure, the potential benefits of utilizing organic compost can be assessed for each study site—Cimone, Markisa, Gedong Pompa, and Cikini Kramat. This risk-informed approach supports the safe and beneficial use of organic compost while guiding targeted waste management practices tailored to local conditions. (Singh et al. 2018).

4.6 Conclusion

Urban kampungs face a significant challenge in waste management due to the high proportion of organic waste and the frequent mixing of household hazardous and toxic waste, which increases the risk of heavy metal contamination in compostable materials. This study evaluates the cancer and non-cancer risks associated with the use of compost derived from urban organic waste in agricultural applications. The findings indicate that heavy metals are present in organic waste, with varying concentrations and types across different kampungs, resulting in distinct health risk profiles. For example, lead (Pb) exposure in Cimone presents a medium cancer risk in 89.84% of cases, a low cancer risk in 10.16%, and a low non-cancer risk in 99.82%. In Markisa, (Pb) exposure results in a medium cancer risk of 93.57%, a low cancer risk of 6.43%, and a low non-cancer risk of 99.53%, while mercury (Hg) exposure in the same location shows a high non-cancer risk in 99.38% of cases. In Gedong Pompa, arsenic (As) exposure leads to a high cancer risk in 69.71% of cases, alongside a low non-cancer risk in 70.89%. Additionally, (Pb) exposure in this area yields a medium cancer risk of 60.89% and a low non-cancer risk of 97.2%. In Cikini Kramat, (Hg) exposure results in a high non-cancer risk in 76.09% of cases. Given these findings, the application of composting in Markisa and Gedong Pompa is not advisable under current conditions due to the elevated non-cancer and cancer risks posed by mercury (Hg) and arsenic (As) exposure, respectively. These varying levels of health risk highlight the importance of incorporating both cancer and non-cancer risk assessments when evaluating compost use in urban agriculture. To mitigate health risks and promote safer composting practices, local governments and communities must strengthen and enhance waste management systems through targeted improvements in infrastructure, education, and policy implementation. A key component of this enhancement include the effective segregation of organic and inorganic waste streams, with particular emphasis on separating household hazardous and toxic waste. Public education and awareness campaigns on safe composting techniques will play a vital role in minimizing heavy metal contamination, thereby fostering safer compost use and promoting sustainable urban agriculture.

Acknowledgements We thank Mr. Nurdin as the community leader of Pasar Baru Sub-district and Mr. Herman as the neighborhood unit leader of Cimone for their support and permission to hold this study in Cimone and Markisa. This research project was supported by the Universitas Indonesia Pintermidi (Inter and Multi-Disciplinary Research) Grant (NKB-3170/UN2.RST/HKP.05.00/2020)

References

Adianto, J., and M. Iwamoto. 2011. Urban Kampung as a part of megacity living system. In *Sensible high density: Megacity design studio Indonesia–Japan 2011*, ed. A. Okabe and H. Ota, 10–36. London: Narifumi Tachimoto (RIHN).

Alizadeh, A., M. Balali-Mood, A. Mahdizadeh and B. Riahi-Zanjani. 2017. Mercury and lead levels in common soaps from local markets in Mashhad, Iran. *Iranian Journal of Toxicology 11*(4), 1–3. http://ijt.arakmu.ac.ir/article-1-598-en.html

Allen, R. G. 1995. *Analysis of heavy metals in cleaning products*. State Agric L Use Baseline.

ASTM International. 2016. *Standard Test Method for Determination of the Composition of Unprocessed Municipal Solid Waste*. [ASTM D5231-92(2016)]. https://doi.org/10.1520/D5231-92R16

ATSDR. 2009. *Toxicological Profile for Cadmium*. Agency Toxic Substance and Disease Registry, U.S. Department of Health and Human Services, Public Health Service.

ATSDR. 2010. *Toxicological Profile for Arsenic*. Agency Toxic Substance and Disease Registry, U.S. Department of Health and Human Services, Public Health Service.

ATSDR. 2018. *Toxicological Profile for Mercury, Vol. 176*. Agency for Toxic Substances and Disease Registry. U.S. Department of Health and Human Services, Public Health Service.

ATSDR. 2019. *Toxicological Profile for Lead* (Draft for Public Comment). Agency for Toxic Substances and Disease Registry, U.S. Department of Health and Human Services, Public Health Service.

Basir-Cyio, M., I. Wahyudi, U. Hasanah, S. Laude, T. Inoue and T. Kawakami. 2018. *Tithonia diversifolia* compost for decreasing the activity of mercury in soil. *Journal of Environmental Science and Technology 11*(2), 79–85. https://www.cabdirect.org/cabdirect/abstract/20183319577

Bello, H. 2018. Impact of changing lifestyle on municipal solid waste generation in residential areas: Case study of Qatar. *International Journal of Waste Resource* 8 (2): 1–7. https://doi.org/10.4172/2252-5211.1000335.

Central Bureau of Statistics. 2019. *Provinsi DKI Jakarta Dalam Angka 2019* [DKI Jakarta Province in Number 2019].

Central of Disease C. 1991. *Preventing lead poisoning in children.* Available in https://stacks.cdc.gov/view/cdc/97650.

Chen, A., K. N. Dietrich, X. Huo, and S. M. Ho. 2011. Developmental neurotoxicants in E-waste: An emerging health concern. *Environmental Health Perspective* 119 (4): 431–438. https://doi.org/10.1289/ehp.1002452.

Davison, A. G., et al. 1988. Cadmium fume inhalation and emphysema. *Lancet* 331 (8587): 663–667. https://doi.org/10.1016/S0140-6736(88)91474-2.

Delgado, O. B., S. Ojeda-Benítez, and L. Márquez-Benavides. 2007. Comparative analysis of hazardous household waste in two Mexican regions. *Waste Management* 27 (6): 792–801. https://doi.org/10.1016/j.wasman.2006.03.022.

Dorne, J., G. Kass, L. Bordajandi, B. Amzal, U. Bertelsen, A. Castoldi, C. Heppner, M. Eskola, S. Fabiansson, P. Ferrari, E. Scaravelli, E. Dogliotti, P. Fuerst, A. Boobis, and P. Verger. 2015. Human risk assessment of heavy metals: Principles and applications. In *Metal ions in toxicology: Effects, interactions, interdependencies*, ed. H. Sigel, A. Sigel, and R. Sigel, 27–60. London: De Gruyter. https://doi.org/10.1515/9783110436624-007.

Dunnivant, F. M., and E. Anders. 2006. *A basic introduction to pollutant fate and transport: An integrated approach with chemistry, modeling, risk assessment, and environmental legislation.* London: Wiley.

Ellisa, E. 2016. Coping with crowding in high-density kampung housing of Jakarta. *Archnet-IJAR: International Journal of Architectural Research 10*(1), 195–212. https://www.archnet.org/publications/10543

Emurotu, J. E., and P. C. Onianwa. 2017. Bioaccumulation of heavy metals in soil and selected food crops cultivated in Kogi State, north central Nigeria. *Environmental System Research* 6 (1): 1–9. https://doi.org/10.1186/s40068-017-0098-1.

Endah, S. R. N., and S. I. Surantaatmadja. 2019. The determination of heavy metals level: Lead in cosmetic soap preparation by atomic absorption spectrophotometer (AAS). *Journal of Physics: Conference Series* 1179 (1): 012178.

Ernst, W. H. O. 1996. Bioavailability of heavy metals and decontamination of soils by plants. *Applied Geochemistry* 11 (1–2): 163–167. https://doi.org/10.1016/0883-2927(95)00040-2.

Esakku, S., K. Palanivelu, and K. Joseph. 2003. Assessment of heavy metals in a municipal solid waste dumpsite. *Workshop on Sustainable Landfill Management* 35: 139–145.

Ferrari, K., R. Gamberini, and B. Rimini. 2016. The waste hierarchy: A strategic, tactical and operational approach for developing countries. The case study of Mozambique. *International Journal of Sustainable Development Plan* 11 (5): 759–770. https://doi.org/10.2495/SDP-V11-N5-759-770.

Fjeld, R. A., N. A. Eisenberg, and K. L. Compton. 2007. *Quantitative environmental risk analysis for human health.* London: Wiley.

Fjelsted, L., and T. H. Christensen. 2007. Household hazardous waste: Composition of paint waste. *Waste Management and Research* 25 (6): 502–509. https://doi.org/10.1177/0734242X07082956.

Fijriah, W. H., and E. Ellisa. 2018. Space of worship in high-density informal settlements: Case study of Kampong Cikini Kramat, Jakarta. *The Journal Social Sciences Research* 6: 591–598. https://doi.org/10.32861/jssr.spi6.591.598.

Flora, C. B., Flora, J. L., Gasteyer, S. P. 2007. *Rural Communities: Legacy and Change.* Boulder, Colorado: Westview Press.

Hendrawan, A. 2012. *Urban Design Studio 2* Kampung *Cikini Kramat dan Kawasan Cikini* [Urban Design Studio 2, Kampung Cikini Kramat and Cikini Areas]. Depok.

Herat, S. 2008. Green electronics through legislation and lead free soldering. *Clean–soil, Air, Water* 36 (2): 145–151. https://doi.org/10.1002/clen.200700164.

IFLS. 2014. *Indonesia family life survey 2014 book 1 (Konsumsi).*

IFLS. 2014. *Indonesia family life survey 2014 book US health measurements.*

Indonesian National Standard. 1994. *Metode pengambilan dan pengukuran contoh timbulan dan komposisi sampah perkotaan* [Methods of taking and measuring samples of municipal solid waste generation and composition] (SNI 19-3964-1994). The National Standardization Agency of Indonesia. https://www.nawasis.org/portal/download/digilib/265-SNI-1994_3964_19.pdf

International Agency for Research on Cancer. 2012. Arsenic, Metals, Fibres, and Dusts: Volume 100c, A Review of Human Carcinogens. *IARC monographs on the evaluation of carcinogenic risks to humans.* https://monographs.iarc.fr/ENG/Monographs/vol100C/mono100C-8.pdf

Isrun, Cyio, M. B., I. Wahyudi, U. Hasanah, S. Laude, T. Inoue, T. Kawakami. 2018. Tithonia diversifolia compost for decreasing the activity of mercury in soil. *Journal of Environmental Science and Technology* 11 (2): 79–85.

Iswanto, I., S. Sudarmadji, E. T. Wahyuni, and A. H. Sutomo. 2016. Timbulan Sampah B3 Rumah-tangga Dan Potensi Dampak Kesehatan Lingkungan Di Kabupaten Sleman, YOGYAKARTA (Generation of Household Hazardous Solid Waste and Potential Impacts on Environmental Health in Sleman Regency, YOGYAKARTA). *Jurnal Manusia Dan Lingkungan* 23 (2): 179–188. https://doi.org/10.22146/jml.18789.

Korzun, E. A., and H. H. Heck. 1990. Sources and fates of lead and cadmium in municipal solid waste. *Journal of the Air and Waste Management Association* 40 (9): 1220–1226. https://doi.org/10.1080/10473289.1990.10466766.

Liu, J., Q. Li, W. Gu, and C. Wang. 2019. The impact of consumption patterns on the generation of municipal solid waste in China: Evidences from provincial data. *International Journal of Environmental Research and Public Health* 16 (10): 1717. https://doi.org/10.3390/ijerph16101717.

Marshall, G., C. Ferreccio, Y. Yuan, M. N. Bates, C. Steinmaus, S. Selvin, et al. 2007. Fifty-year study of lung and bladder cancer mortality in Chile related to arsenic in drinking water. *JNCI Journal of the National Cancer Institute* 99 (12): 920–928. https://doi.org/10.1093/jnci/djm004.

Mascagni, P., D. Consonni, G. Bregante, G. Chiappino, and F. Toffoletto. 2003. Olfactory function in workers exposed to moderate airborne cadmium levels. *Neurotoxicology* 24 (4–5): 717–724. https://doi.org/10.1016/S0161-813X(03)00024-X.

Masebinu, S. O., E. T. Akinlabi, E. Muzenda, A. O. Aboyade, C. Mbohwa, M. M. Manyuchi. and P. Naidoo. 2017. A review on factors affecting municipal solid waste generation. *Paper presented at 2nd International Engineering Conference 2017, Minna, Nigeria.* http://hdl.handle.net/10210/250378

Masindi, V. and K. L. Muedi. 2016. Environmental contamination by heavy metals. In: Saleh, H. E. M. and R. F. Aglan, editors. *Heavy metals.* 117–134. https://doi.org/10.5772/intechopen.76082

Mohee, R., and N. Soobhany. 2014. Comparison of heavy metals content in compost against vermi-compost of organic solid waste: Past and present. *Resource, Conservation and Recycling* 92: 206–213. https://doi.org/10.1016/j.resconrec.2014.07.004.

Monavari, S. M., G. A. Omrani, A. Karbassi, and F. F. Raof. 2012. The effects of socioeconomic parameters on household solid-waste generation and composition in developing countries (a

case study: Ahvaz, Iran). *Environmental Monitoring and Assessment* 184: 1841–1846. https://doi.org/10.1007/s10661-011-2082-y.
Mooney, C. Z. 1997. *Monte Carlo simulation.* Thousand Oaks: Sage Publications.
Morales, K. H., L. Ryan, T. L. Kuo, M. M. Wu, and C. J. Chen. 2000. Risk of internal cancers from arsenic in drinking water. *Environmental Health Perspective* 108 (7): 655–661. https://doi.org/10.1289/ehp.00108655.
Mulyadin, R. M., K. Ariawan, and M. Iqbal. 2018. Conflict of waste management in DKI Jakarta and its recomended solutions. *Jurnal Analisis Kebijakan Kehutanan* 15 (2): 179–191. https://doi.org/10.20886/jakk.2018.15.2.179-191.
Nelles, M., J. Grünes, and G. Morscheck. 2016. Waste management in Germany—Development to a sustainable circular economy? *Procedia Environmental Sciences* 35: 6–14. https://doi.org/10.1016/j.proenv.2016.07.001.
OEHHA. 2007. *Lead and lead compounds.* California Office of Environmental Health Hazard Assessment. https://oehha.ca.gov/chemicals/lead-and-lead-compounds
Ojeda-Benítez, S., Q. Aguilar-Virgen, P. Taboada-González, and S. E. Cruz-Sotelo. 2013. Household hazardous wastes as a potential source of pollution: A generation study. *Waste Management and Research* 31 (12): 1279–1284. https://doi.org/10.1177/0734242X13510057.
Ong, C. N., W. O. Phoon, H. Y. Law, C. Y. Tye, H. H. Lim. 1985. Concentrations of lead in maternal blood, cord blood, and breast milk. *Archives of Disease in Childhood* 60 (8): 756–9. August. https://doi.org/10.1136/adc.60.8.756. PMID: 4037861; PMCID: PMC1777426.
Page, A. L., T. J. Logan, and J. A. Ryan. 1987. *Land application of sludge: Food chain implications.* London: Lewis Publishers.
Pinamonti, F., G. Stringari, F. Gasperi, and G. Zorzi. 1997. The use of compost: Its effects on heavy metal levels in soil and plants. *Resources, Conservation and Recycling* 21 (2): 129–143. https://doi.org/10.1016/S0921-3449(97)00032-3.
Randall, P. M. 1995. Mercury reduction in products and processes: A review of the electrical and electronic industries. *Environmental Progress* 14 (4): 232–239. https://doi.org/10.1002/ep.670140411.
Rosesar, J. S., Kristanto, G. A. 2018. Household solid waste composition and characterization in Indonesia Urban Kampong. *IOP Conference Series: Materials Science and Engineering* 909 (1), 012077.
Saidalavi, R., A. Hashim, K. B. Kishor, P. K. Leena, and P. Adake. 2017. Analysis of lead and arsenic in cosmetics and assessment of students awareness about cosmetic toxicity. *International Journal of Basic and Clinical Pharmacology* 6 (6): 2319–2003. https://doi.org/10.18203/2319-2003.ijbcp20172235.
Singh, J., R. Saxena, V. Bharti, and A. Singh. 2018. The importance of waste management to environmental sanitation: A review. *Advances in Bioresearch* 9 (2): 202–207. https://doi.org/10.15515/abr.0976-4585.9.2.202207.
Smolinska, B. 2015. Green waste compost as an amendment during induced phytoextraction of mercury-contaminated soil. *Environmental Science and Pollution Research* 22 (5): 3528–3537. https://doi.org/10.1007/s11356-014-3601-5.
Tangerang Central Bureau of Statistics. 2018a. *Jumlah Penduduk menurut Kecamatan dan Jenis Kelamin di Kota Tangerang* [Total Population by District and Gender in Tangerang City].
Tangerang Central Bureau of Statistics. 2018b. *Profil Sanitasi Kota Tangerang Vol. 5* [Tangerang City Sanitation Profile].
Tchounwou, P. B., Yedjou, C. G., Patlolla, A. K., Sutton, D. J. 2012. Heavy metal toxicity and the environment. Molecular, Clinical, and Environmental Toxicology. *Experimentia Supplementum* 101: 133–164. https://doi.org/10.1007/978-3-7643-8340-4_6. PMID: 22945569; PMCID: PMC4144270.
Tchobanoglous, G., and F. Kreith. 2002. *Handbook of solid waste management. Handbook of sustainability for the food sciences*, 2nd ed. London: McGraw Hill Handbooks.

U.S. Environmental Protection Agency. 1995. *Mercury, elemental; CASRN 7439-97-6*. Integrated Risk Information System, National Center for Environmental Assessment. https://cfpub.epa.gov/ncea/iris/iris_documents/documents/subst/0370_summary.pdf
U.S. Environmental Protection Agency. 2000. *Notice of filing a pesticide petition to establish a tolerance for a certain pesticide chemical in or on food*. Federal Register.
Wang, G., H. Q. Liu, Y. Gong, Y. Wei, A. J. Miao, L. Y. Yang, et al. 2017. Risk assessment of metals in urban soils from a typical Industrial city, Suzhou, Eastern China. *International Journal of Environmental Research and Public Health* 14 (9): 1–6. https://doi.org/10.3390/ijerph14091025.
Wang, Q., D. Zhou, L. Cang, L. Li, and H. Zhu. 2009. Indication of soil heavy metal pollution with earthworms and soil microbial biomass carbon in the vicinity of an abandoned copper mine in Eastern Nanjing, China. *European Journal of Soil Biology* 45 (3): 229–234. https://doi.org/10.1016/j.ejsobi.2008.12.002.
Whittle, A. J., and A. J. Dyson. 2002. The fate of heavy metals in green waste composting. *The Environmentalist* 22 (1): 13–21. https://doi.org/10.1023/A:1014563807198.
Yasuda, K., and M. Tanaka. 2006. Report on hazardous household waste generation in Japan. *Waste Management Research: the Journal for a Sustainable Circular Economy* 24 (4): 397–401. https://doi.org/10.1177/0734242X06064821.
Zhou, H., W. T. Yang, X. Zhou, L. Liu, J. F. Gu, W. L. Wang, et al. 2016. Accumulation of heavy metals in vegetable species planted in contaminated soils and the health risk assessment. *International Journal of Environmental Research and Public Health* 13 (3): 289. https://doi.org/10.3390/ijerph13030289.

Chapter 5
Investigation of Microbial Air Quality in Urban Kampung Adjacent to Landfill

Ova Candra Dewi, Zakianis, Nadia Adrianti, Intan Chairunnisa, and Baskoro Laksitoadi

Abstract This chapter investigates the microbial air quality of one kampung adjacent to a landfill. Kampung Nambo, with its steep topography, is located less than 300 m from Cipeucang landfill. The landfill has been operated since 2011 and received municipal solid waste from Tangerang Selatan City. The waste decomposition at the landfill produces odors and germs that spread all over the kampung and beyond. Seventy-four houses out of 153 houses in Kampung Nambo were tested. Further investigation is done through a four-quadrant matrix chart and linear regression using the house's orientation to the landfill, distance from the landfill, elevation compared to the landfill, and cross ventilation as the variables. Outdoor and indoor air flow was simulated through SimScale's computational fluid dynamic (CFD) modeling software for visualizing the air movement. Finding shows that 67% of the houses have air quality exceeding the limit of the allowed threshold. Secondly, distance from the landfill has the most correlation and impact to the microbial air quality, followed by humidity, orientation, elevation, density, and lastly cross ventilation. This study found that cross ventilation is not easily applicable in houses adjacent to a landfill due to the bad outdoor air quality. Furthermore, barriers such as walls and trees are proven to reduce the level of air pollution in the house.

Keywords Airflow · Airborne diseases · House openings · Landfill · Microbial air quality

O. C. Dewi (✉) · N. Adrianti · I. Chairunnisa · B. Laksitoadi
Faculty of Engineering, Universitas Indonesia, Depok, Indonesia
e-mail: ova.candewi@ui.ac.id

I. Chairunnisa
e-mail: intanisac@ui.ac.id

B. Laksitoadi
e-mail: baskorolaksi@ui.ac.id

Zakianis
Faculty of Public Health, Universitas Indonesia, Depok, Indonesia
e-mail: zakianis.arifin@gmail.com

M. Budianta et al. (eds.), *A Liveable Kampung*, Engaging Indonesia,
https://doi.org/10.1007/978-981-96-7866-2_5

5.1 Introduction

Most of the air pollution mitigation related to urban studies took place in cities with high rise buildings and wide streets (Yang et al. 2020; Carpentieri and Robins 2015; Miao et al. 2020). Nevertheless, major air pollution also happened in the peri-urban areas where landfills are located and the microbial contamination occurred (Basinska et al. 2019; Cyprowski et al. 2019). The existence of microbial contaminants in the air can cause irritation in the lungs, breathing problems, allergies, fever, and pneumonia (Pastuszka et al. 2000). Many researchers have proof that landfills potentially decreased the surrounding air quality (Liu et al. 2016). Unfortunately, there are not many studies exposing the microbial air quality in the settlement adjacent to the landfill, including in Indonesia. The World Bank estimated about 68% of the world's population will live in urban areas by 2050 (Our World in Data 2019). It leads to an increasing demand for land conversion, including for landfill needs. Land extensive is one of the factors which makes landfill becoming unsustainable (Kulkarni 2020), moreover the odors and germs it brought.

Air pollution, as defined by WHO, is contamination of the indoor or outdoor environment by any chemical, physical, or biological agent that modifies the natural characteristics of the atmosphere (World Health Organization 2021). Air pollution, which can be potentially caused by the landfill, spread over the settlement in the vicinity by the microbial contamination factor, such as bacteria, fungi, and mold (Basinska et al. 2019; Cyprowski et al. 2019). Although air pollution can be controlled through urban vegetation, not many plant types can be used for biological filters and the environment might be too harsh for said plants to live (Przybysz et al. 2020). It can also absorb considerable amounts of trace elements.

Kampung Nambo, South Tangerang, is located adjacent to the Cipeucang Landfill. This is one example of kampungs which is located adjacent to a landfill in Indonesia. The landfill has been operated since 2011 and served the population of South Tangerang City. The city is a fast-growing city and currently it has 3.6 million inhabitants. Every day, the landfill received around 970 tons of mix waste, which contained 49.7% organic fraction (DLH Kota Tangerang Selatan 2014). In 2018–2019, roughly there were more than 250 cases of pneumonia, and more than 200 cases of TBC identified in the district where the landfill is located (Puskesmas Serpong). Preliminary test of indoor microbial air quality was taken in October 2019 and resulted in 67% of the housing samples exceeding the limit of microbial quality. However, it is surprising that in the area less than 300 m from the landfill, 33% of the houses were having a good quality of air. This result triggers further study on investigating the microbial air quality in Kampung Nambo from the perspective of airflow analysis. It sets up actions for improving the indoor air quality for houses located adjacent to a landfill and recommends policies for future landfill site selection and improves microbial air quality for kampung adjacent to a landfill.

5.2 Literature Review

5.2.1 Urban Kampung Adjacent to a Landfill

Urban kampung is defined as a settlement in an urban area with a certain community that is vulnerable and adaptable at the same time (Putri and Herlily 2020). These settlements are often built informally and spontaneously. A study by Ernawati et al. (2013) stated that kampung may deal with challenges from the city development (Putri and Herlily 2020). Waste related issues are one of the challenges that are faced by many urban kampungs in big cities (Rosesar and Kristanto 2020). Moreover, along with the increasing population of the rapidly growing city, the need for waste disposal areas has also increased. When waste disposal areas are not optimized, it can result in polluting air, water, land, and lastly harming human health (Rosesar and Kristanto 2020).

A study in Western Australia has identified potential health impacts from the expansion of a sanitary landfill next to a proposed housing site (Little et al. 2007). Moreover, In Campania Region, South Italy, serious issues with Municipal Waste Management have caused several public disturbances especially in odor and environmental pollution issues (Feo et al. 2013). This odor related issue also occurs in Kampung Nambo, Indonesia, which is carried by the wind from Cipeucang landfill (Dewi et al. 2019). Referring to previous examples, there is a big possibility that improper landfill disposal may cause disturbance to urban settlement whether to its comfort, health, and/or its environment.

5.2.2 Indoor Air Quality in Kampung Adjacent to Landfill

Many studies discuss the effect of wind in the distribution of air pollution in an urban settlement. Creating a good air flow in an urban environment to some extent can reduce the time near-surface pollutants stagnate and prevent the accumulation of air pollutants, which also improves air quality (Yang et al. 2020), but it should also be noted that wind may distribute air pollution, largely affected by its speed and direction (Łowicki 2019; Epstein et al. 2017; Yang et al. 2020). Wind velocity and direction will change as the near-surface urban spatial conditions change, which can affect near-surface air pollutant dispersion (Epstein et al. 2017). This means spatial form has direct effects on the wind environment which includes wind velocities, directions, and near-surface turbulences, and indirect effects on air pollution through the wind environment (Lee 2019). The influence of urban spatial forms and wind environment should be included in consideration of near-surface urban air pollution dispersion (Yang et al. 2020). Thus, it should be noted that no matter how favored natural ventilation is in humid climates like Indonesia, for a settlement adjacent to landfill, natural ventilation potentially becomes the starting point of indoor air pollution.

In this chapter, there are six variables investigated that might affect the indoor air quality:

1. Cross ventilation velocity
2. Distance from the landfill
3. Elevation compared to the landfill
4. Orientation compared to the landfill
5. Humidity
6. Occupant density

5.2.3 Cross Ventilation Velocity

According to local regulation, the ideal indoor air speed is around 0.15–0.25 m/s (Peraturan Menteri Kesehatan Republik Indonesia 1077/Menkes/ Per/V/2011). In addition, SNI 03-1733-2004 mentions that the speed can be added as temperature rises to reach comfort (Badan Standardisasi Nasional 2004).

5.2.4 Distance to Landfill

Distance from the settlement to the landfill affects people's perception of environmental consequences. The nearer the houses to the landfill, the more chances of decreasing in residents' concerns about odor-associated health issues (Feo et al. 2013). Thus, for housing that is located adjacent to landfill, an air buffer zone should be added to maintain the air quality around the houses. This buffer zone may be in the form of a certain distance from the landfill to the settlement. Concluded from several local regulations in several countries, this distance varies in range of 100–500 m. In Indonesia, the Ministry of Public Works has determined the buffer zone from the landfill should be at least 500 m (Peraturan Menteri Pekerjaan Umum No. 19/PRT/M 2012).

A comparison of several buffer zone distance requirements in various countries is provided in Table 5.1.

Indonesia has the same buffer zone for each landfill category that has been regulated by law, which is at least 500 m or more. The risk factors considered for the buffer zone are leachate infiltration into springs, the danger of methane gas explosion, and the spread of disease vectors (Peraturan Menteri Pekerjaan Umum Nomor 19/PRT/M 2012). The region that does not specify their buffer zone, like the United States, states that the buffer zone must be determined through environmental study (Criteria for Solid Waste Disposal Facilities EPA 1993).

Table 5.1 Zone distance requirement

Country	Climate	Buffer zone	Source
Canada (British Columbia Region)	4 Seasons	Not specified	Central Pollution Control Board India, 2017
Hong Kong	4 Seasons	Has two type of buffer zone: dusty landfill needs buffer zone at least 100–200 m and landfill with odor needs buffer zone at least 200 m	Extension of Existing Landfills and Identification of Potential New Waste Disposal Sites, 2003
		The documents also stated that the lowest boundary is 100 m and upper boundary is 500 m	
South Africa	4 Seasons	At least 500 m	Central Pollution Control Board India, 2017
USA	4 Seasons	Not specified	Criteria for Solid Waste Disposal Facilities EPA, 1993
Iran	4 Seasons	At least 500 m	Mahboobeh Derakhshandeh, Toktam Taleb Beydokhti, 2014
Europe	4 Seasons	Not specified	European Parliamentary: Official Journal C 156, 24/05/1997 P. 0010, 1997
Australia	4 Seasons	Not specified	BC Ministry of Environment, 2017
India	6 Seasons	At least 100–500 m	Central Pollution Control Board, 2017
Bangladesh	Monsoon, 3 Seasons	At least 250 m	Central Pollution Control Board India, 2017
Indonesia	Tropical	At least 500 m	Minister of Public Works Regulation no. 19 of 2012
Malaysia	Tropical	At least 500 m	Central Pollution Control Board India, 2017
Tropic countries according to ISWA	Tropical	Not specified	ISWA (International Solid Waste Association), 2013

Table 5.2 Official standard of air quality, density, and humidity

No.	Variable	Standard	Source
1	Microbial air quality	700 CFU/m	Peraturan Menteri Kesehatan Republik Indonesia Nomor 1077/MENKES/PER/V/2011
2	Density	6.4–9.6 m^2/adult	Keputusan Kepala Badan Standardisasi Nomor 30/KEP/BSN/11/2004
		3.2–4.8 m^2/children	
3	Humidity levels	40–60% Rh	Peraturan Menteri Kesehatan Republik Indonesia 1077/Menkes/Per/V/2011

5.2.5 House Elevation and Orientation Compared to the Landfill

Houses that have lower elevation compared to the landfill are not favored because of the high probability of a valley breeze during the day that brings the germs into the house. In terms of orientation, houses that face the landfill are not favored because it has a higher probability to be infiltrated by germs when wind blows from the direction of the landfill (Feo et al. 2013).

5.2.6 Occupant Density

The minimum space requirement at the house is defined by relating the need for air space needs for activities that are carried out at home such as sleeping, cooking, taking bath, sitting, etc. With such activities, the need for normal air space for adults is between 16 and 24 m^3 per hour, while children need 8–12 m^3 air space per hour (Baiche and Walliman 2002; Badan Standardisasi Nasional 2004).

If it is translated to the number of an area, it means that with a room height $\geq$ 2.5 m, the minimum area for each adult is 6.4–9.6 m^2, and 3.2–4.8 m^2 for a child (Baiche and Walliman 2002). However, it is added that the volume of space per person for living rooms can be reduced to 7.5 m^3, and bedrooms to 10 m^3 per bed if the room is facilitated with ducting or natural ventilation (Baiche and Walliman 2002). Therefore, the minimum land area needed for a house can be calculated by considering the number of family members, and not forgetting the Building Coverage Ratio which is specified in local regulations (Badan Standardisasi Nasional 2004).

5.2.7 Humidity

Room humidity should be maintained in the range of 40–70% with the healthiest range of 40–60% and the most pleasant range in 50–60% (Badan Standardisasi Nasional 2001; Baiche and Walliman 2002; Peraturan Menteri Kesehatan Republik Indonesia 1077/Menkes/Per/V/2011). This concludes that the humidity of the air should be in the range of 50%-60% to maintain healthy and pleasant air.

5.2.8 Improving Air Quality in Houses Adjacent to Landfill

Improvement of urban ventilation can be considered as one of the effective methods to reduce the pollutant exposure of near-road residents. Unlike cities in western countries, peri-urban in developing country cities has an organic spatial form. In addition, buildings in peri-urban are non-high-rise residential buildings. Sometimes roads and buildings are on different levels (Ellisa 2016; Rutz 1987).

Natural ventilation through window opening is also used as a way to get fresh air (Heiselberg et al. 2001). However, airflow is needed to make the fresh air enter the room. Air moves when there is a pressure difference (Baiche and Walliman 2002). Not only that, direction of air entering windows, size, details of windows, interior partitioning details, and location also take part as factors to determine the pattern of airflow (Lechner 2015). Opened windows that are located on opposite walls are the most ideal configuration as it will allow cross ventilation that is desired in humid climates, however, the indoor airflow itself depends on the air pressure distribution within the room (Lechner 2015).

Unfortunately, natural ventilation may actually contaminate indoor air quality in an area that has polluted air. To resolve that, ensuring enough distance for the buffer zone is needed. This buffer zone may also be complemented with vegetation. Green landscape planning can be applied if there's enough available land/space (Łowicki 2019). This and other studies show that the effect of vegetation varies depending on particle sizes, distance to the sources, emission level, wind speed, precipitation to humidity ratio, and so on. Moreover, installing exhaust fans that may extract pollution before entering the building is also applicable when eliminating pollution sources is not doable (Lechner 2015).

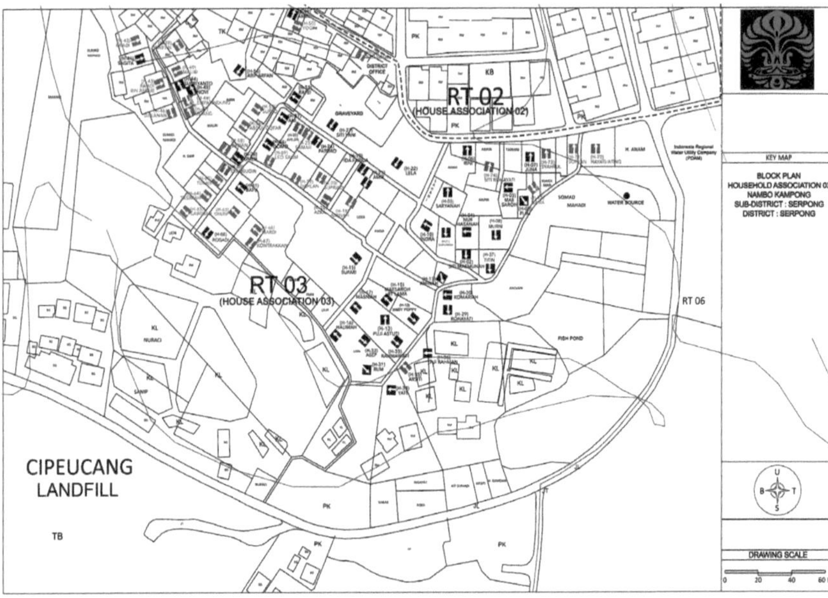

Fig. 5.1 Site plan of Kampung Nambo (drawn by Universitas Indonesia's research team)

5.3 Methods

5.3.1 Site Overview

Data collection in Kampung Nambo is initially planned to be carried out in 153 houses. But on the site, only 75 house owners agree to partake in the research. To define the houses orientation, Fig. 5.1 was drawn by the research team through field visit and the use Google Earth. Kampung Nambo is located side-by-side with the landfill (Figs. 5.2, 5.3).

5.3.2 Data Collection

In the 64 houses, we collect microbial air quality data through bacteria and fungi collection methods. Microbial data for bacteria and fungi using a water vacuum device, petri dishes for fungi (malt extract agar) and bacteria (trypticase soy agar), aluminum foil, and alcohol. Air intakes are carried out sequentially starting with the first bacterial data collection. Microbial air intake is carried out in two places for each house; the first one is in the living room, assuming that the room is a room that is often used by homeowners and outside the house. The microbial air intake in the living room needs 120 s and 180 s in the outside of the house. During the air intake,

Fig. 5.2 Landfill in the background (personal collection)

Fig. 5.3 Landfill surrounding the houses (personal collection)

the researchers take data on room temperature, air velocity, and humidity. The petri dishes that have been filled with bacteria and fungi are then taken to the Universitas Indonesia's Engineering Faculty laboratory for incubation.

The microbial air result is then compared to the healthy microbial air standard. Houses are divided into two, the ones below healthy microbial air standard and above healthy microbial air standard. From there, the research group was divided into 4 groups based on the site elevation.

Data collection is then continued to 2 steps:

1. Physical data collection for Kampung Nambo in the form of: house plan, orientation of the house to the landfill, distance of the house to the landfill, elevation. The house plan is then redrawn as the basis for simulating the airborne simulation scenario.
2. Taking questionnaires to homeowners about house openings, homeowner habits related to house openings' usage and the house openings' condition itself.

The collected data is then compared to the standard of each variable:

5.3.3 Data Analysis

Outdoor and indoor airflow is simulated through SimScale computational fluid dynamic (CFD) modeling software for visualizing the air movement that comes from the landfill. Inlet speed for the simulation is 3.5 m/s determined by using Beaufort Scale on the survey. Indoor cross ventilation simulated with the scenario where all windows are closed but the doors are open. This scenario is based on the survey and interview conducted previously. It is revealed that most of the houses in Kampung Nambo rarely open their windows because of the bad odor in the neighborhood (Figs. 5.4, 5.5).

The results from the simulation and other data samples are then compared with the air quality data in the form of a four-quadrant matrix, with the variables as the *X* value and microbial air quality as the *Y* value. Quadrant I is the ideal condition where variables and microbial air quality values are according to the standard regulation.

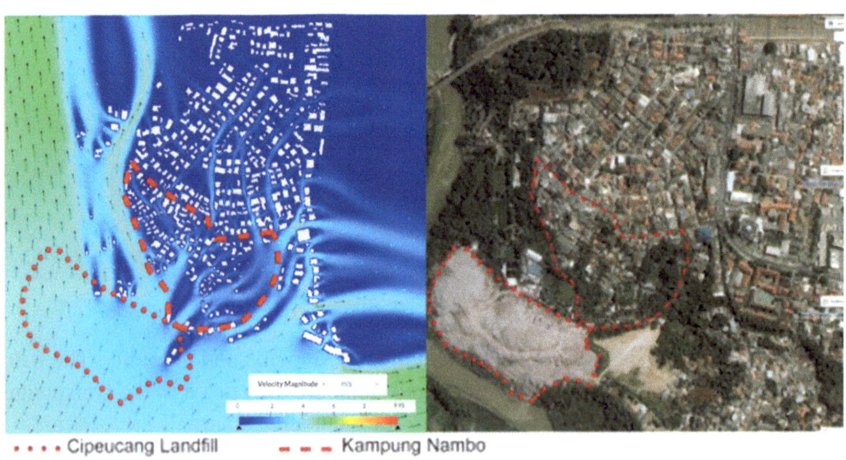

Fig. 5.4 Cipeucang landfill from above (combination of SimScale CFD modeling software and Google Earth)

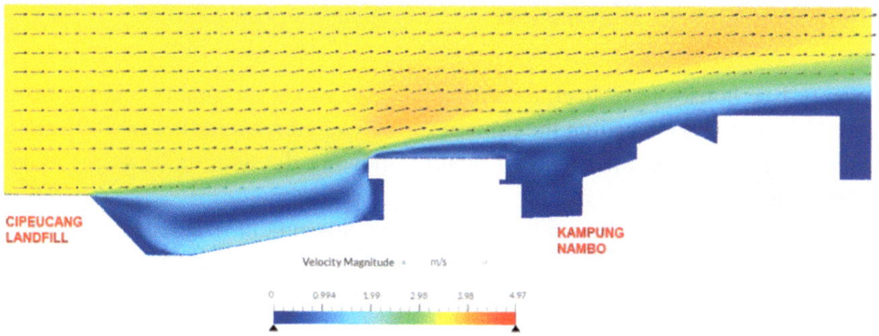

Fig. 5.5 Air quality simulation (SimScale CFD modeling software)

Quadrant II where variables are according to standard, but the air quality is not, Quadrant III where variables are not according to standard but the air quality is good. Finally, Quadrant IV where both variables and air quality are not according to the standard regulation (Fig. 5.6).

On the last part, linear regression analysis is conducted for each variable to find which variable has the highest or lowest impact on the microbial air quality. Using the value of Multiple R, we can determine how strong the correlation of the 6 variables is to the microbial air quality data (0 = no correlation, 1 = strong correlation).

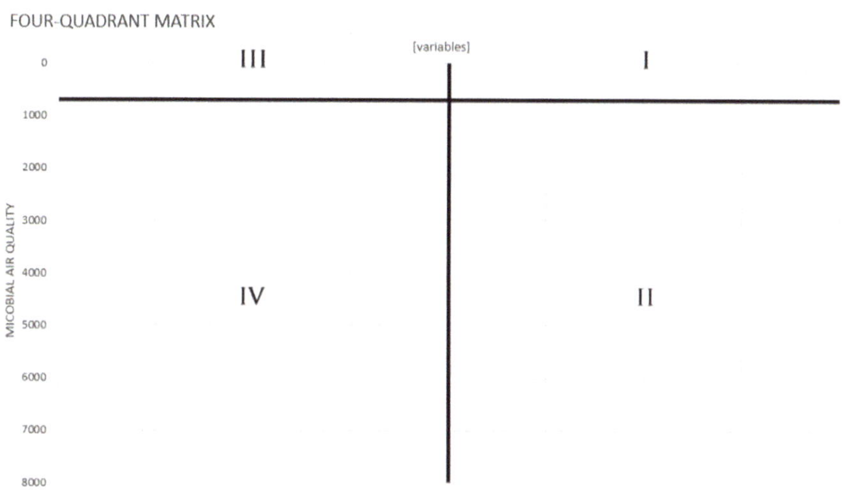

Fig. 5.6 Four-quadrant matrix

5.4 Results and Discussion

Seventy-five houses have been tested for their microbial air quality and the results show 75% of them exceeding the limit of the allowed threshold (below 700). Below is the summary of the microbial air quality, showing the number of germs in each house (Table 5.3).

From the CFD simulation, 63% of the sample does not have a decent cross ventilation velocity. While on the other hand, the average cross ventilation velocity is 0.925 m/s for the houses that have (Figs. 5.7, 5.8, 5.9).

5.4.1 *Four-Quadrant Matrix*

According to the Peraturan Menteri Kesehatan Republik Indonesia Nomor 1077/MENKES/PER/V/2011, a healthy indoor environment has a minimum cross ventilation velocity at 0.15 m/s. Negative number means that there was turbulent flow inside the house, but cross ventilation did not occur (Figs. 5.10, 5.11).

According to the Peraturan Menteri Pekerjaan Umum No. 19/PRT/M/2012, a healthy house should have a minimum distance of 500 m from the landfill (Figs. 5.12, 5.13).

Table 5.3 Summary of the microbial air quality

Sample	Germs	Sample	Germs	Sample	Germs	Sample	Germs	Sample	Germs
H-01	**282**	**H-02**	**142**	**H-03**	**142**	**H-04**	**142**	**H-05**	**141**
H-06	**106**	**H-07**	**142**	**H-08**	**212**	**H-09**	**142**	H-10	7279
H-11	2720	H-12	2509	**H-13**	**142**	H-14	1413	H-15	3568
H-16	2650	H-17	6289	H-18	1166	H-19	1696	H-20	1519
H-21	5406	H-22	2367	H-23	1025	H-24	2544	H-25	1025
H-26	1944	H-27	5088	H-28	3039	H-29	3180	H-30	3710
H-31	2332	H-32	2968	H-33	2156	H-34	3428	H-35	1590
H-36	4806	H-37	4488	H-38	3533	H-39	706	H-40	954
H-41	**212**	H-42	1237	H-43	1060	**H-44**	**353**	H-45	2120
H-46	883	H-47	778	H-48	848	**H-49**	**283**	H-50	812
H-51	813	**H-52**	**212**	H-53	1025	**H-54**	**636**	H-55	778
H-56	**530**	**H-57**	**566**	H-58	813	H-59	1131	H-60	1660
H-61	**566**	H-62	919	H-63	1025	H-64	1166	H-65	1130
H-66	1130	H-67	1448	**H-68**	**672**	H-69	1166	H-70	1166
H-71	1060	H-72	1837	H-73	1590	H-74	1731	H-75	1307

*Bold: below the threshold (700)

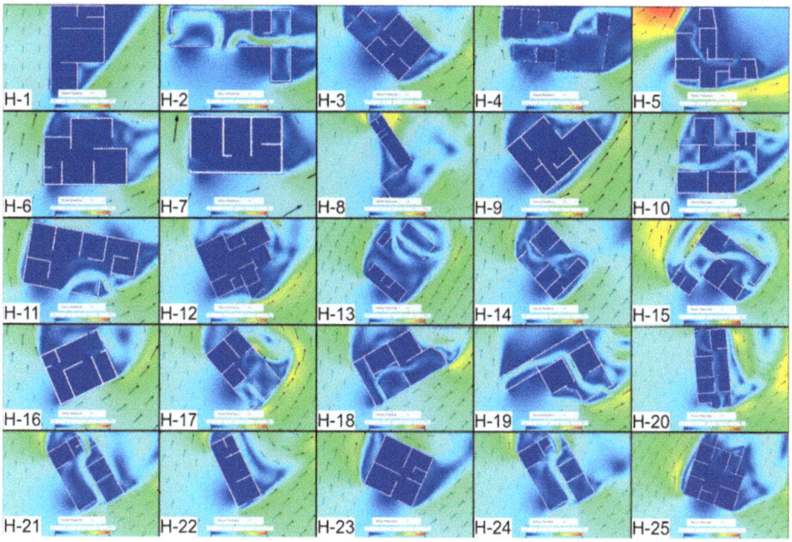

Fig. 5.7 CFD simulation of house 1–25 (SimScale CFD modeling software)

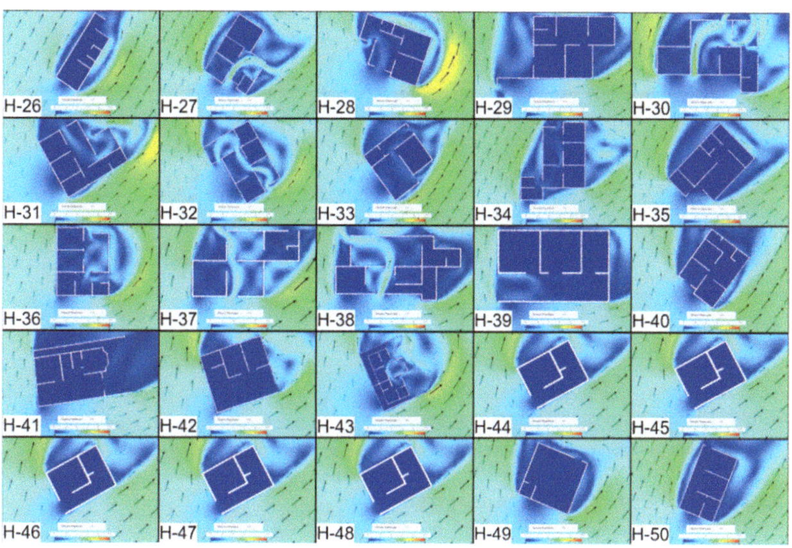

Fig. 5.8 CFD simulation of house 26–50 (SimScale CFD modeling software)

Houses that have lower elevation compared to the landfill are not favored because of the high probability of a valley breeze during the day. From the survey, landfill elevation is 115 feet above sea level. Houses below that elevation are not favorable. And in terms of orientation, houses that face the landfill are not favored because it

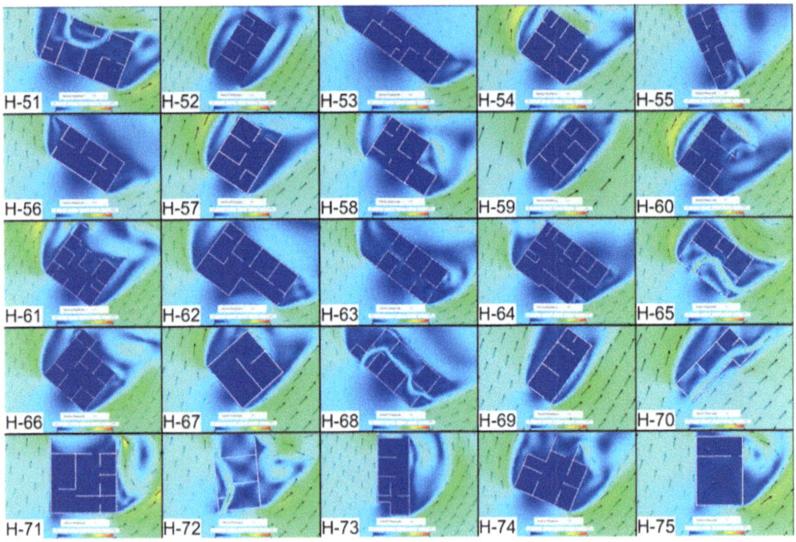

Fig. 5.9 CFD simulation of house 51–70 (SimScale CFD modeling software)

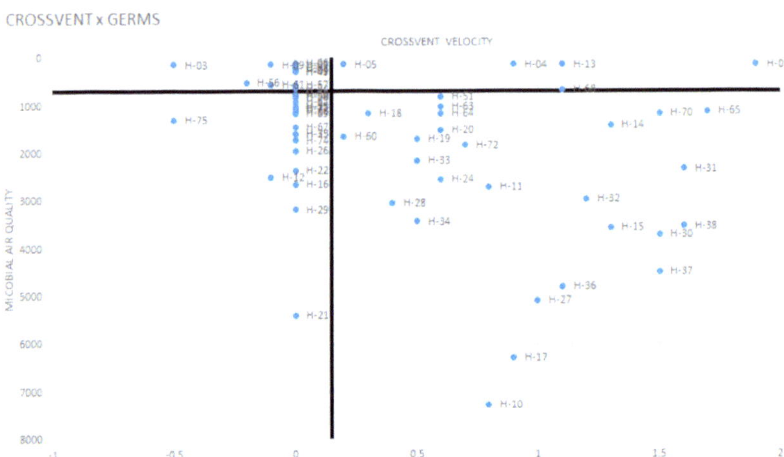

Fig. 5.10 Matrix of cross ventilation and germs

has a higher probability to be infiltrated by the germs when a wind blows from the direction of the landfill (Feo et al. 2013) (Fig. 5.14).

According to the Peraturan Menteri Kesehatan Republik Indonesia Nomor 1077/MENKES/PER/V/2011, a healthy indoor environment has humidity value between 40 and 60% (Fig. 5.15).

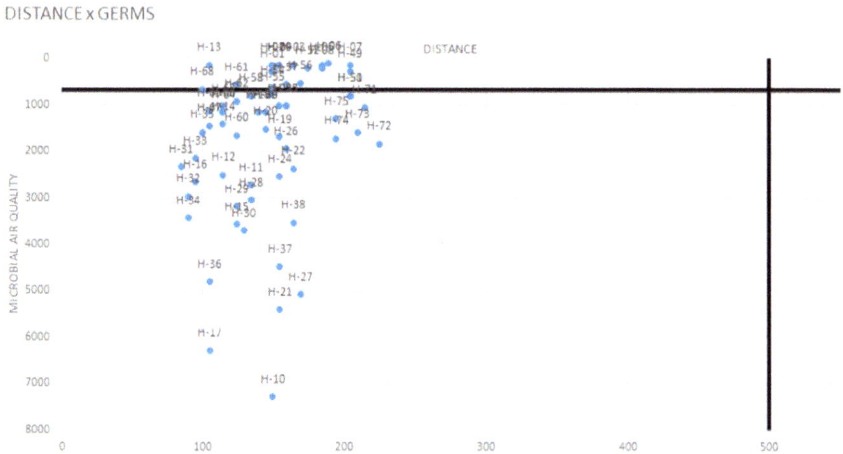

Fig. 5.11 Matrix of distance and germs

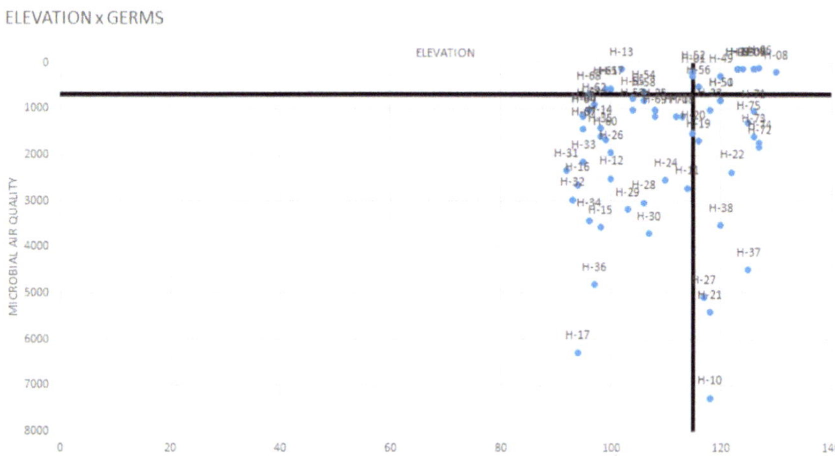

Fig. 5.12 Matrix of elevation and germs

According to the SNI 03-1733-2004, the minimum floor area for an adult in a healthy house is 9.6 m^2 per person.

A summary of the four-quadrant matrix for each sample is compiled in the diagram below (Fig. 5.16).

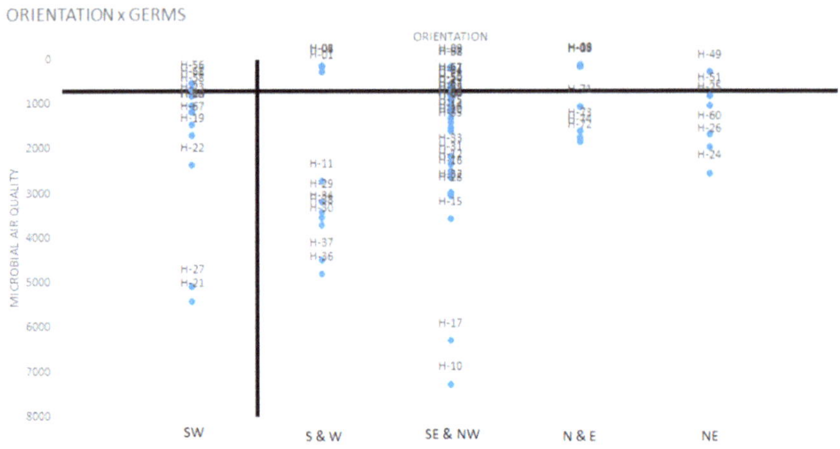

Fig. 5.13 Matrix of orientation and germs

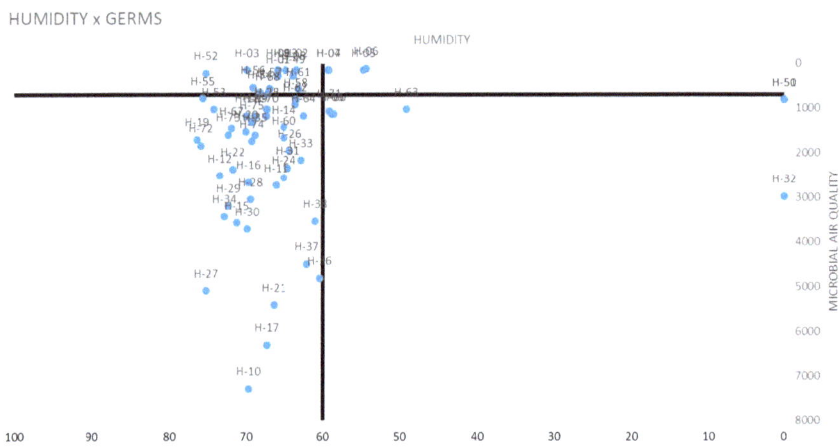

Fig. 5.14 Matrix of humidity and germs

5.4.2 *Linear Regression*

	Cross vent	Distance	Elevation	Orientation	Humidity	Density
Multiple *R*	0.160072	0.522209	0.229942	0.193977	0.442687	0.111098

Based on the Multiple R value, it is shown that distance from the landfill has the strongest correlation to the microbial air quality, followed by humidity, elevation, orientation, cross ventilation, and, last, occupant density.

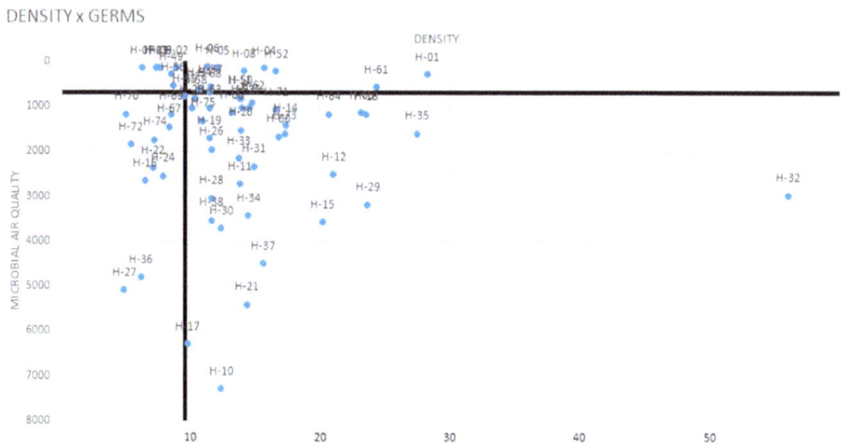

Fig. 5.15 Matrix of density and germs

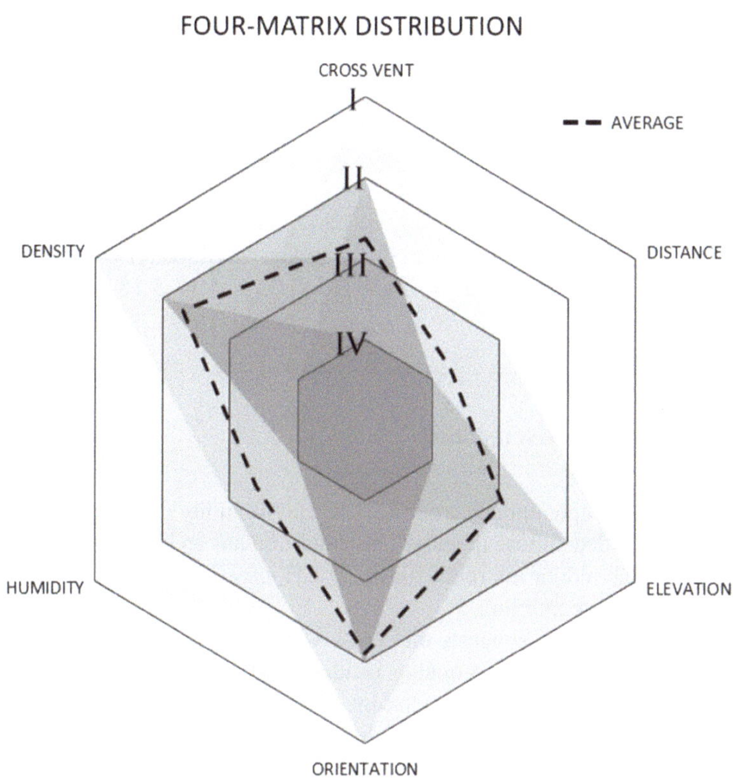

Fig. 5.16 Four matrix contribution

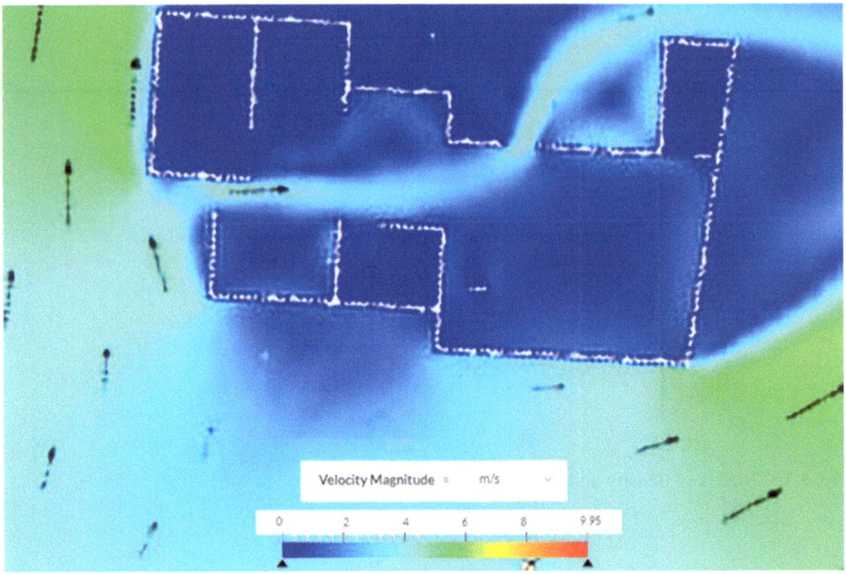

Fig. 5.17 H-04 CFD simulation (SimScale CFD modeling software)

5.5 Discussion

Further investigation of 2 samples with the best ideal condition (H-04 and H-05) and 2 others with the worst condition (H-67 and H-69) (Figs. 5.17, 5.18, 5.19, 5.20, 5.21, 5.22, 5.23, 5.24).

5.6 Results and Discussion

With an ambient temperature of around 30 °C and humidity of 80%, organic waste at the landfill will decompose quickly. This condition has an impact on air pollution and at the end, the indoor microbial air quality. Houses adjacent to the landfill will be affected with this condition.

While some literature suggests that cross ventilation is a factor that has to be fulfilled to create a healthy air quality inside a house, this study suggests that it is not easily applicable in houses adjacent to a landfill. Cross ventilation has the lowest impact to create a healthy air quality when there is a source of pollution in the neighborhood.

Some steps need to be considered to increase the indoor air quality such as inlet that comes from clean places; increasing the total air flow to the crowded space; maximizing laminar flow (preferably vertical) rather than turbulent; directing the

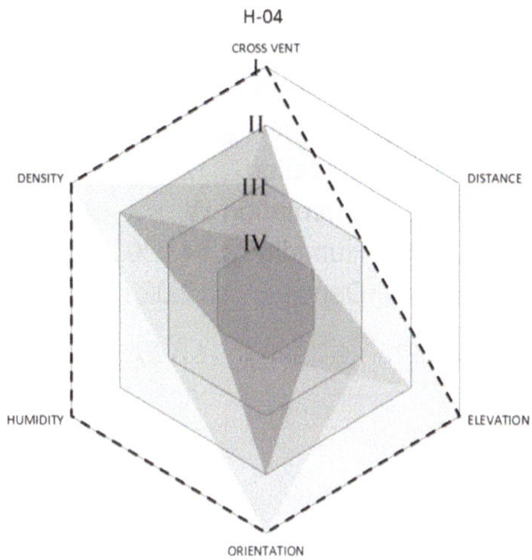

H-04.

Air quality: 142 CFU/m
Cross Vent : 0.9 m/s
Distance : 115 m
Elevation : 126 fasl
Orientation : West
Humidity : 59.2%
Density : 15.69 m^2/person

Fig. 5.18 Result of H-04 overall condition

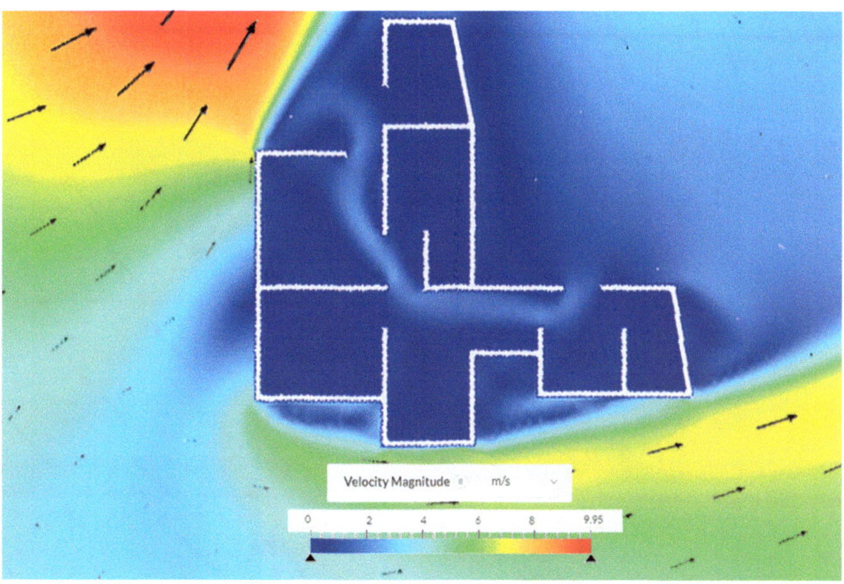

Fig. 5.19 H-05 CFD simulation (SimScale CFD modeling software)

flow of potentially contaminated air out of the room and away from people; ensuring comfort and stable airspeed (Balgeman et al. 2020a, b).

While the location of the house cannot be changed (distance and elevation factors), this study suggests that existing houses should avoid having an inlet that faces the landfill directly. Whether they move its main entrance or never open existing inlets

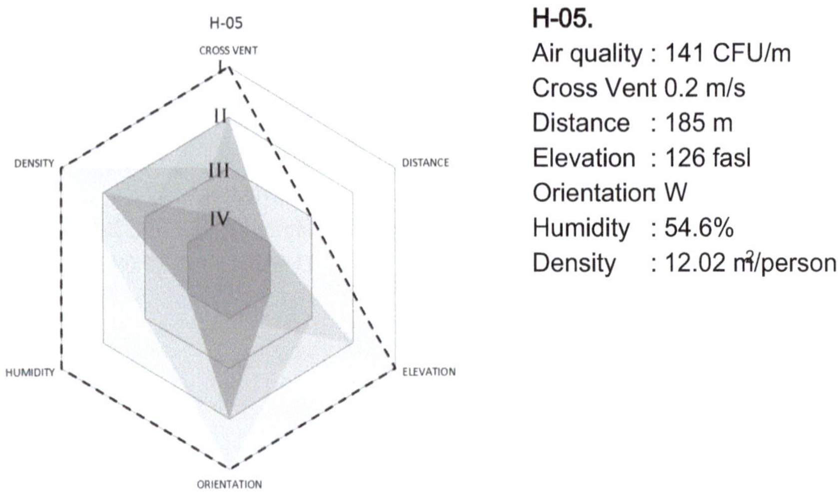

Fig. 5.20 H-05 overall condition

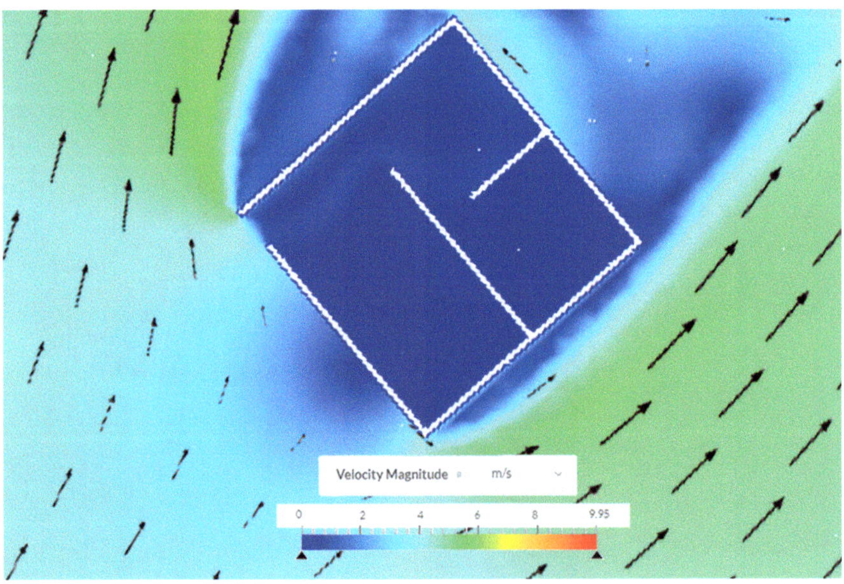

Fig. 5.21 H-67 CFD simulation (SimScale CFD modeling software)

(doors and windows). Furthermore, barriers such as walls and trees are proven to reduce the level of air pollution in the house.

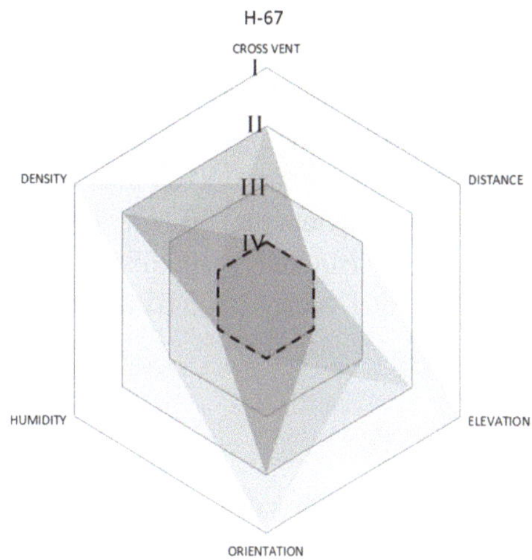

H-67.

Air quality : 1448 CFU/m
Cross Vent : 0 m/s
Distance : 105 m
Elevation : 95 fasl
Orientation : SW
Humidity : 71.85%
Density : 8.34 m^2/person

Fig. 5.22 H-67 overall condition

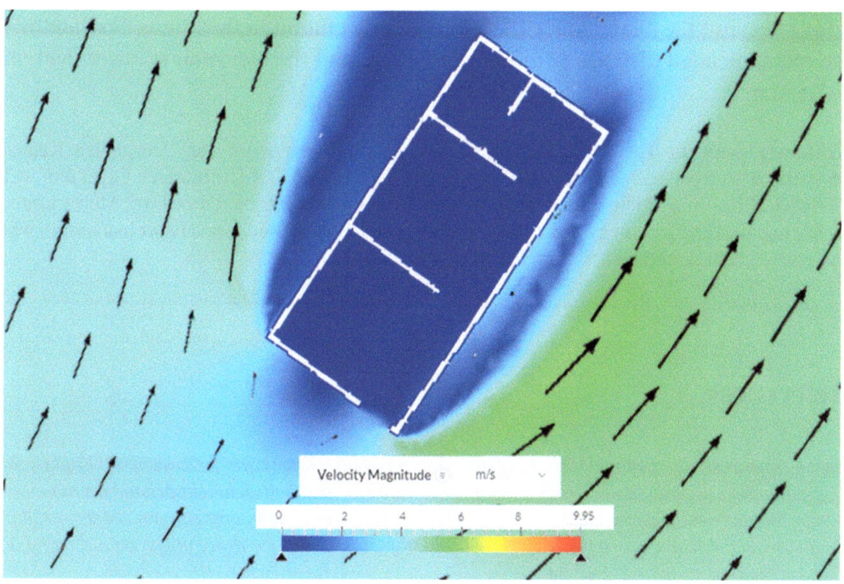

Fig. 5.23 H-69 CFD simulation (SimScale CFD modeling software)

5.7 Conclusion

This study suggests that existing houses should avoid facades/openings facing the landfill directly. Facades/openings facing the landfill must be closed and the orientation of the facade of the house changed. Boundaries such as walls and trees can be

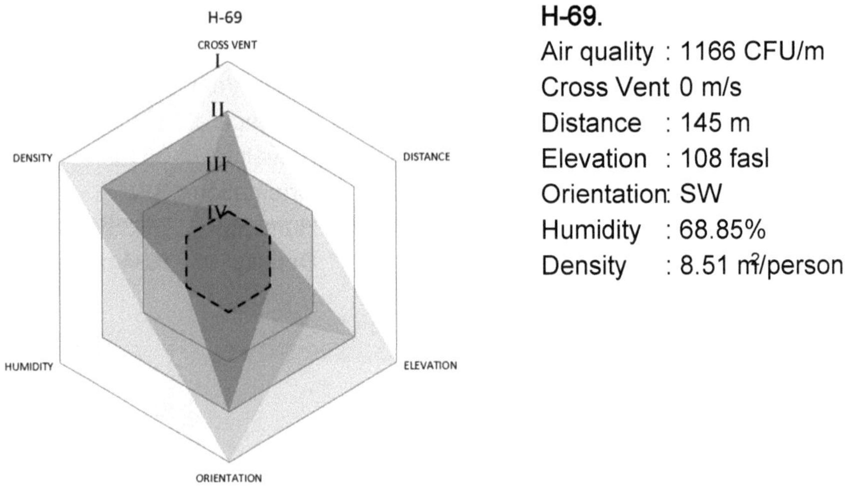

Fig. 5.24 H-69 overall condition

added because they can reduce the level of air pollution in the house. For landfills, the recommended distance is more than 500 m, and the elevation is lower than the settlement.

Acknowledgements This study was supported by Direktorat Riset dan Pengabdian Kepada Masyarakat Universitas Indonesia (DRPM UI) under PINTERMIDI UI Research Grant No. 945/SK/R/UI/2019. This article cannot be completed without the contribution of Prof. Melva Louisa (Department of Pharmacology and Therapeutics, Faculty of Medicine, Universitas Indonesia), who had given feedback and input during the writing process

References

Badan Standarisasi Nasional. 2004a. SNI-03-1733-2004. *Tata cara perencanaan lingkungan perumahan di perkotaan (Urban planning concepts for residential neighborhood areas).*

Badan Standarisasi Nasional. 2004b. SNI 03-6572-2001. *Tata cara perancangan sistem ventilasi dan pengkondisian udara pada bangunan gedung (Procedures for designing ventilation and air conditioning systems in buildings).*

Balgeman, S., B. Meigs, S. Mohr, S. Niemoller and P. Spranzi. 2020. *Can HVAC systems help prevent transmission of COVID-19?* Advanced Industries Practice, McKinsey & Company. https://www.mckinsey.com/industries/industrials-and-electronics/our-insights/can-hvac-systems-help-prevent-transmission-of-covid-19/

Basińska, M. Ł, M. Ł Michałkiewicz, and K. Ratajczak. 2019. Impact of physical and microbiological parameters on proper indoor air quality in nursery. *Environment International* 132: 105098. https://doi.org/10.1016/j.envint.2019.105098.

British Columbia Ministry of Environment. 2016. *Landfill criteria for municipal solid waste 2016.*

Carpentieri, M., and A. G. Robins. 2015. Influence of urban morphology on air flow over building arrays. *Journal of Wind Engineering and Industrial Aerodynamics* 145: 61–74. https://doi.org/10.1016/j.jweia.2015.06.001.

Central Pollution Control Board (CPCB). 2017. *Guidelines on the provision of buffer zone around waste processing and disposal facilities Central Pollution Control Board*. 1–24.

Cyprowski, M., A. Ławniczek-Wałczyk, M. Ł Gołofit-Szymczak, K. Frączek, J. Kozdrój, and R. ŁL. Górny. 2019. Bacterial aerosols in a municipal landfill environment. *Science of the Total Environment* 660: 288–296. https://doi.org/10.1016/j.scitotenv.2018.12.356.

De Feo, G., S. De Gisi, and I. D. Williams. 2013. Public perception of odour and environmental pollution attributed to MSW treatment and disposal facilities: A case study. *Waste Management* 33 (4): 974–987. https://doi.org/10.1016/j.wasman.2012.12.016.

Derakhshandeh, M., and T. T. Beydokhti. 2014. Management of landfill locating of urban waste. *European Online Journal of Natural and Social Sciences* 3 (3): 32–39.

Dewi, O. C., T. U. Atmoko, H. A. Alrasyid, D. Priyandana, M. B. V. Dewi, and D. Siska. 2019. Tackling environmental pollution through participatory approach in Nambo, South Tangerang, Indonesia. *IOP Conference Series: Earth and Environmental Science* 291: 012003. https://doi.org/10.1088/1755-1315/291/1/012003.

Ellisa, E. 2016. Coping With Crowding in High-density Kampung Housing of Jakarta. *International Journal of Architectural Research: Archnet-IJAR* 10 (1): 195. https://doi.org/10.26687/archnet-ijar.v10i1.790.

Environmental Protection Department. 2003. *Extension of existing landfills and identification of potential new waste disposal sites final strategic environmental assessment report*. 1(January).

Epstein, S. A., S.-M. Lee, A. S. Katzenstein, M. Carreras-Sospedra, X. Zhang, S. C. Farina, P. Vahmani, P. M. Fine, and G. Ban-Weiss. 2017. Air-quality implications of widespread adoption of cool roofs on ozone and particulate matter in southern California. *Proceedings of the National Academy of Sciences* 114 (34): 8991–8996. https://doi.org/10.1073/pnas.1703560114.

Ernawati, R. 2013. Facing urban vulnerability through Kampung development, case study of Kampungs in Surabaya. *Humanities and Social Sciences* 1 (1): 1–6. https://doi.org/10.11648/j.hss.20130101.11.

European Parliament. 1997. *Proposal for a COUNCIL DIRECTIVE on the landfill of waste*, Official Journal C 156, 24/05/1997 P. 0010.

Heiselberg, P., K. Svidt, and P. V. Nielsen. 2001. Characteristics of airflow from open windows. *Building and Environment* 36 (7): 859–869. https://doi.org/10.1016/s0360-1323(01)00012-9.

Keputusan Kepala Badan Standardisasi Nomor 30. 2004. *Penetapan pedoman standardisasi nasional nomor 302-2004 sertifikasi produk-pedoman umum pelaksanaan*. (*Issuance of guidelines for national standardization No 302-2004 product certification—General guidelines for implementation*). Jakarta.

Kulkarni, B. N., and V. Anantharama. 2020. Repercussions of COVID-19 pandemic on municipal solid waste management: Challenges and opportunities. *Science of the Total Environment* 743: 140693. https://doi.org/10.1016/j.scitotenv.2020.140693.

Lechner, N. 2015. *Heating, cooling, lighting: Sustainable design methods for architects*, 4th ed. London: Wiley.

Lee, C. 2019. Impacts of urban form on air quality: Emissions on the road and concentrations in the US metropolitan areas. *Journal of Environmental Management* 246: 192–202. https://doi.org/10.1016/j.jenvman.2019.05.146.

Little, M., Earnest, J., Spickett, J., Karscherian, D. (2007). The role of stakeholders in HIA: A Landfill Site and Housing Development in Mundijong, Western Australia. *Environmental Health* 7(4), pp. 65–73. http://hdl.handle.net/20.500.11937/20423

Liu, Y., W. Lu, H. Guo, Z. Ming, C. Wang, S. Xu, Y. Liu, and H. Wang. 2016. Aromatic compound emissions from municipal solid waste landfill: Emission factors and their impact on air pollution. *Atmospheric Environment* 139: 205–213. https://doi.org/10.1016/j.atmosenv.2016.05.043.

Łowicki, D. 2019. Landscape pattern as an indicator of urban air pollution of particulate matter in Poland. *Ecological Indicators* 97: 17–24. https://doi.org/10.1016/j.ecolind.2018.09.050.

Miao, C., S. Yu, Y. Hu, R. Bu, L. Qi, X. He, and W. Chen. 2020. How the morphology of urban street canyons affects suspended particulate matter concentration at the pedestrian level: An in-situ investigation. *Sustainable Cities and Society*. https://doi.org/10.1016/j.scs.2020.102042.
Munawar, E. and J. Fellner. 2013. *Guidelines for design and operation of municipal solid waste landfills in tropical climates. The international solid waste association report.*
Neufert, P., B. Baiche, and N. Walliman. 2002. *Architects' data*. London: Wiley.
Peraturan Menteri Kesehatan Republik Indonesia Nomor 1077. 2011. *Pedoman penyehatan udara dalam ruang rumah*. Menteri Kesehatan Republik Indonesia. Jakarta. (*Guidelines for household air quality improvement*. Jakarta, Ministry of Health, Republic of Indonesia).
Pastuszka, J. S., U. K. T. Paw, D. O. Lis, A. Wlasto, and K. Ulfig. 2000. Bacterial and fungal aerosol in indoor environment in Upper Silesia, Poland. *Atmospheric Environment* 34 (22): 3833–3842. https://doi.org/10.1016/S1352-2310(99)00527-0.
Peraturan Menteri Pekerjaan Umum Nomor 19. 2012. *Pedoman penataan ruang kawasan sekitar tempat pemrosesan akhir sampah. Menteri Pekerjaan umum.* Jakarta. (*Guidelines for spatial arrangement of areas near the final waste treatment processing*. Ministry of Public Work, Jakarta).
Przybysz, A., M. Wińska-Krysiak, M. Małecka-Przybysz, M. Stankiewicz-Kosyl, M. Skwara, A. Kłos, S. Kowalczyk, K. Jarocka, and P. Sikorski. 2020. Urban wastelands: On the frontline between air pollution sources and residential areas. *Science of the Total Environment* 721: 137695. https://doi.org/10.1016/j.scitotenv.2020.137695.
Putri, N. S., and Herlily. 2020. Investigating community adaptability and resilience in Urban Kampung, study case: Water and sanitation infrastructure of Kampung Muka, North Jakarta. *IOP Conference Series: Earth and Environmental Science* 452: 012095. https://doi.org/10.1088/1755-1315/452/1/012095.
Rutz, W. 1987. *Cities and Towns in Indonesia: Their Development, Current Positions and Functions With Regard to Administration and Regional Economy (Urbanization O)* (English and German Edition). Lubrecht & Cramer Ltd.
Semilu Rosesar, J., and G. Andari Kristanto. 2020. Household solid waste composition and characterization in Indonesia Urban Kampong. *IOP Conference Series: Materials Science and Engineering* 909: 012077. https://doi.org/10.1088/1757-899x/909/1/012077.
SimScale. 2025. SimScale Modelling Computational Fluid Dynamics. [Computer Software]. SimScale, GmbH. https://www.simscale.com/
The Advanced Industries Practice, Balgeman, S. B., Meigs, B., Mohr, S., Niemöller, A., & Spranzi, P. 2020, July. *Can HVAC systems help prevent the transmission of COVID-19?* Mckinsey and Company. https://www.mckinsey.com/~/media/McKinsey/Industries/Advanced%20Electronics/Our%20Insights/Can%20HVAC%20systems%20help%20prevent%20transmission%20of%20COVID19/Can-HVAC-systems-help-prevent-transmission-of-COVID-19-final.pdf
United States Environmental Protection Agency. 1993. *Criteria for solid waste disposal facilities a guide for owners/operators*. (March), 11–13.
World Health Organization. 2021. Air pollution. https://www.who.int/health-topics/air-pollution#tab=tab_1
Yang, J., B. Shi, Y. Zheng, Y. Shi, and G. Xia. 2020. Urban form and air pollution disperse: Key indexes and mitigation strategies. *Sustainable Cities and Society* 57: 101955. https://doi.org/10.1016/j.scs.2019.101955.

Chapter 6
Herbal Medicine Usage in Urbanized Kampungs and Peri-urban Communities

Manneke Budiman, Bantari W. K. Wardhani, Halvina Saiya, Dhita Hapsarani, and Rista Ihwanny

Abstract The impact of urbanization on kampung and peri-urban communities is often far-reaching and total. It touches almost all of the sectors of the communities' everyday life. However, some communities have demonstrated their resilience toward certain aspects of the modernization process that is brought about by urbanization. The article aims at examining the local communities' resilience in using herbal medicines for self-medication despite the continuous process of modernization as the result of urbanization in three urban kampungs in Jakarta and Tangerang. The long history of self-medication practice using herbal medicines was not only caused by herbal medicines' accessibility and low cost but also because such a practice serves as the communities' means of negotiation with modern medicine, which represents modernity, a key element of urbanization. Using a combination of ethnographic research methods and questionnaire distribution to collect data, this article suggests that there is a correlation between the communities' cultural history and their attitude toward herbal medicines.

M. Budiman (✉) · D. Hapsarani · R. Ihwanny
Department of Literature, Faculty of Humanities, Universitas Indonesia, Depok, Indonesia
e-mail: manneke.budiman@ui.ac.id

D. Hapsarani
e-mail: dhita@ui.ac.id

R. Ihwanny
e-mail: rista.ihwanny@umn.ac.id

B. W. K. Wardhani
Research Center for Pharmaceutical Ingredients & Traditional Medicine, Badan Riset & Inovasi Nasional, Jakarta, Indonesia
e-mail: bant001@brin.go.id

H. Saiya
School of Environmental Science, Universitas Indonesia, Jakarta Pusat, Indonesia
e-mail: halvina.grasela@ui.ac.id

R. Ihwanny
Film & Animasi, Universitas Multimedia Nusantara, Tangerang, Banten, Indonesia

M. Budianta et al. (eds.), *A Liveable Kampung*, Engaging Indonesia,
https://doi.org/10.1007/978-981-96-7866-2_6

Keywords Community resilience · Cultural history · Herbal medicine · Self-medication · Urban kampung

6.1 Introduction

A rapid and massive expansion of urban areas to the surrounding peripheries of cities has been most prominent in major cities in the developing world. The suburban and peri-urban areas that used to have visible rural characteristics have continually been exposed to significant changes as a result of large-scale urbanization, and the unique rural dimensions of their communities have been disappearing at an alarming rate. Similarly, urban kampung areas, which are commonly characterized by a high population density, limited water resources, poor waste management, high unemployment rate, and unhygienic lifestyle, have also been more severely exposed to the ceaseless onslaught of urbanization. It causes their living environment to shrink further and their quality of life to degrade.

Much of the literature on these issues addresses the loss of local practices and traditions that the peri-urban communities and urban kampungs suffer from the expanding city space. While in general the existing studies discuss the persistence of herbal medicine utilization practices in these two types of urban communities, very few, if any, look deeper into the socio-political motives that underlie such persistence. Most of the reasons put forward by the existing studies are concerned with the transgenerational continuation of transmitted traditions and practices. Lubis, for instance, pointed out that many Indonesians across ethnic groups still strongly believed in traditional herbal medicine as part of their ancestral heritage, and that herbal medicine was developed based on local wisdom and knowledge to address societal needs (2009, p. 535). Nevertheless, there is an underlying assumption that the communities choose to continue the tradition for the sake of tradition or based on the tacit knowledge of herbal medicine that they are morally obliged to preserve without any well-informed understanding of the composition of the medicine or its therapeutic effect.

This article's standpoint is different in the sense that it tries to establish a correlation between the persistence of herbal medicine usage among urban kampung and peri-urban communities on one hand and the communities' strategy of negotiating with modernity that is brought to them by urbanization. At the same time, they also have to cope with the policies and regulations concerning the utilization of herbal medicine set by health authorities. The article argues that such a negotiation is essential to the communities' survival in terms of the preservation of their local social and cultural identities. Thus, while they cannot put a halt to the invading force of modernity, they still attempt to leave some room for the continuation of their traditions and identities.

The area of Indonesian tropical forests covers about 143 million hectares and is home to about 80% of the world's medicinal plants. It is estimated that the Indonesian tropical forests contain 28,000 plant species (Elfahmi et al. 2014, p. 52). Indonesia, in addition to Brazil, is one of the world's top two centers of biodiversity. Indonesia also houses the largest biodiversity in the world with around 30.000 types of plants, and part of those plants are sources of medicinal materials (Lubis 2009, p. 533). The Indonesian Country Study on Biological Diversity (ICSBD 1992) listed between 25,000 and 30,000 species of flowering plants in Indonesia, with some 10 percent of the total flora of Indonesia being thought to have medicinal value. Furthermore, approximately 40 million Indonesians rely directly on biodiversity, and communities across the archipelago make use of around 6000 plant species on a day-to-day basis (Bappenas 2003, qtd in Lubis 2009, p. 533).

Several studies related to the usage of herbal medicines in Indonesia have been conducted. Nurhayati and Widowati (2017) examined the use of herbal medicines in households in Indonesia in 2013. Pengpid and Peltzer (2018) examined the use of herbal medicines in 2014–2015 in Java. Kautsar, Ayunovani, and Surahman specifically researched the use of herbal medicines in West Java, known as the province that became the largest herbal producer in Indonesia. Meanwhile, Nurhayati and Widowati (2017) conducted a study aimed at identifying dominant factors related to the utilization of traditional health services in households in Indonesia. Their research used household data from Riskesdas (National Basic Health Research) in 2013. The number of subjects analyzed was 294,959. They found that households that live in urban areas have low levels of education, are private workers, have high economic status, know the availability of health services, and are more likely to take advantage of traditional health services.

Furthermore, Kautsar, Ayunovani, and Surahman (2016) examined variables that influence the decision to purchase herbal medicines in Indonesia. Using a cross-sectional method, a survey was conducted on 150 residents in West Java, a province known as the largest herbal medicine producer. The results found that 70% of subjects used herbal medicine, 20% used standardized herbal medicine, and 1.3% used *phytopharmaca*. The social system and characteristics of herbal medicine show a big influence on purchasing decisions of herbal medicines, while the demographics and communication system do not have a big influence.

Research related to traditional medicine usage in rural areas in other countries has been extensively conducted. Maung Maung et al. (2019) examined the use of traditional and modern medicines in the rural region of Kedah, Malaysia. This research aimed at digging into the local community's knowledge about traditional and modern medicines, as well as their preferences in terms of each type of the health issues that they faced. It is qualitative research using FGDs to gather the primary data. The participants of the FGDs were mostly housewives who were over 18 years old. The result showed that modern medicine remains the main preference of the respondents. They also had a solid understanding of the advantages and disadvantages of both traditional and modern medicines. An interesting finding of this research is concerned with the influence of religious belief on the therapy for health issues that they faced. The respondents chose traditional medicine for dealing with mental health which, in their view, should be done based on the Islamic teachings. This research offered an alternative regarding our understanding of traditional medicine usage in a sense that cultural and social background was not the only factor in the use of traditional medicine, but we need to also take into account the religious factor behind such traditional medicine usage.

Zakaria and Zainal (2017), in their article, discussed the construction of the cultural hegemony related to the practice of the Malay traditional medicine, which developed and adapted itself to the Singaporean official medicine. This research used historical archives, surveys on the use of traditional medicine, semi-structured interviews, and in-depth interviews on twelve Malay traditional medicine practitioners, two doctors, and tourism paramedics. Similar to what Maung Maung, Deborah, and Tun wrote, Zakaria and Zainal also mentioned the relationship between Malay traditional medicine and Islam in terms of both the medication practice and medicinal products. The traditional medicinal practice was also believed to have a healing power on mental disorders in comparison to hospital treatment.

Furthermore, the writers also traced the historical archives about the history of Malay traditional medicine related to the animistic belief in necromancy. Using a Gramscian perspective, the writers argued that such a culture-based alternative or traditional medicine seek to find a balance and coexistence rather than a counter-hegemonic act against the Singaporean official medicine. Interestingly, the Malay traditional medicine had an inferior status vis-à-vis Chinese medicine, which has been legally recognized and practiced in hospitals alongside modern medicine by the Singapore government. The important finding related to Malay traditional medicine discussed in this article can be approached from two perspectives. First, medicine hegemony was actually an extension of the modern state, which uses such practice as a political means of controlling its citizens and, second, the Malay/indigenous traditional medicine, through its secondary status, served to complement that hegemony while at the same time challenging.

The historical dimension is the novelty offered by our research. Not only did we use surveys as the source of data, but we also carried out an ethnographic activity by interviewing village elders. Thus, this study explores the correlation between herbal medicine usage and its history. This study examines the use of herbal medicines in four urban kampungs in the Jakarta Greater Capital Region, namely Cimone, Markisa, Cikini, and Gedong Pompa. All of the four urban kampungs have similar characteristics, which include high population density, limited water resources, poor waste management, high unemployment rate, and unhygienic lifestyle.

6.2 Method

A cross-sectional study was conducted using questionnaire distribution to 236 community members in the four research sites from November 2019 to February 2020 which, in addition to the general demographic information, also included information regarding knowledge, perception, and attitude in the use of herbal medicine. The data collected was then input into the SPSS (Statistical Package for the Social Sciences), cleaned, and interpreted. In combination with questionnaires, an ethnographic means for further data collection was also conducted through a series of interviews with specific community leaders and elderlies who were assumed to possess knowledge of the history of herbal medicine practices in their communities, as well as how local social and cultural identities preserved the usage of traditional medicine in urbanization.

6.3 Results

The following tables are the results of the field work in the four research locations, which consist of (1) the demographic information, (2) traditional and herbal medicinal practices, (3) attitude toward traditional and herbal medicine, (4) level of knowledge of traditional and herbal medicines, and (5) types of traditional and herbal medicines consumed by the kampungs' inhabitants. The tables present the dynamics of traditional and modern medicines used, which varies from kampung to kampung based on the locations of the kampungs vis-à-vis the city center as well as the distance from the city center.

The majority of the respondents who participated in the survey are women, with the education backgrounds ranging mostly from elementary education to high school. Most of them were either housewives or self-employed, but fortunately most of them were insured, which is probably due to the fact that all citizens have to join the universal healthcare program (BPJS). The data shown on Table 6.1 may indicate an overlap between the universal healthcare participation and the mandatory one, which are actually the same.

There is no direct correlation between the respondents' low to middle-level education and the fact that most of them were insured with the continuing practice of using traditional and herbal medicines. The key factors that influenced the reasons for consuming herbal medicine of respondents from one kampung to another (and modern medicine, for that matter), as well as the kind of herbal plants that they grew, are shown in the following table.

Table 6.2 shows that the majority of the respondents consumed herbal medicine, but mostly for mild illnesses, and the knowledge about herb remedies was from senior family members (parents). A small percentage of the respondents were aware of the side effects of herbal medicine and also claimed to have experienced the side effects. Most of the herbal plants that they grew in their yards are plants that also happen to be used for cooking and food, including some fruit plants, such as papaya, lime, and soursop. The data also reveal some interesting information, notably that many of the respondents consumed both traditional and modern medicines when ill or, in fact, preferred modern medicine to the traditional one, particularly in Gedong Pompa, where modern medicines could be easily obtained without prescription. A further detailed analysis on this is presented in the Discussion section that follows.

Tables 6.3 and 6.4 provide data on the respondents' attitude toward traditional and herbal medicines, as well as the level of knowledge about them that the respondents possessed in order to make informed and well-thought decisions for consuming herbal medicines when ill. It is clear that they did not follow transmitted tradition

Table 6.1 Demographic table of respondents

Variables	Number (%)
Sex	
Male	34 (14.4)
Female	202 (85.6)
Age	
25–34	29 (12.3)
35–44	68 (28.8)
45–54	66 (28)
55–64	57 (24.2)
$\geq$ 65	16 (6.8)
Education levels	
No elementary education	16 (6.8)
Elementary school	70 (29.7)
Secondary school	69 (29.2)
High school/vocational school	73 (30.9)
Higher education	8 (3.4)
Occupation	
Unemployed	12 (5.1)
Civil Servants	2 (0.8)
Self-employed	24 (10.2)
Housewives	170 (72)
Private sector employees	12 (5.1)
Factory workers	6 (2.5)
Others	10 (4.2)
Health insurance	
Insured	222 (94.1)
Mandatory healthcare	196 (83.1)
Personal healthcare	22 (9.3)
Jamsostek	3 (1.3)
Private insurance	1 (0.4)
Uninsured	14 (5.9)

and practices blindly, for they were aware of the efficacy level and side effects of using traditional and herbal medicines. Thus, the continued use of herbal medicines does not seem to have originated from any superstitious belief or lack of adequate knowledge about the advantages and disadvantages of herbal remedies.

Table 6.5 shows the wide range of types of herbal medicine consumed by the respondents in the four research locations, from the most common herbs such as turmeric, ginger, tamarind, lemon grass, and sage leaf, to the distinct local "herbs" such as soursop leaf and mangosteen skin. Many of these herbs are actually wild

Table 6.2 Traditional and herbal medicinal practices

Variable	Number (%)			
	Markisa	Cimone innovation	Cikini Kramat	Gedong Pompa
Have you ever consumed traditional medicine?				
Yes	39 (83)	40 (72.7)	54 (77.1)	38 (59.4)
No	8 (17)	15 (27.3)	16 (22.9)	26 (40.6)
How often do you use traditional medicine in the last twelve months?				
Once	4 (10.3)	6 (15)	10 (18.5)	4 (10.5)
2 times	3 (7.7)	3 (7.5)	3 (5.6)	2 (5.3)
3 times	4 (10.3)	2 (5)	5 (9.3)	4 (10.5)
> 3 times	28 (71.8)	29 (72.5)	36 (66.7)	28 (73.7)
What is your reason for choosing traditional medicine?				
Mild illnesses	25 (64.1)	20 (50)	22 (40.7)	20 (52.6)
Previous medical history	1 (2.6)	2 (5)	6 (11.1)	10 (26.3)
More economical and practical	8 (20.5)	10 (25)	13 (24.1)	7 (18.4)
Limited access to health facilities	2 (5.1)	1 (2.5)	–	1 (2.6)
Familiarity with the efficacy of the traditional medicine	23 (59)	21 (52.5)	25 (46.3)	17 (44.7)
Low confidence in health officials	–	1 (2.5)	–	1 (2.6)
Others				
Other people's suggestions	1 (2.6)	1 (2.5)	3 (5.6)	–
Maintaining fitness	2 (5.1)	1 (2.5)	4 (7.4)	1 (2.6)
Diet	–	1 (2.5)	1 (1.9)	
More natural and safer	2 (5.1)	3 (7.5)	3 (5.6)	1 (2.6)
Have you ever suffered from the side effect of traditional medicine?				
Yes	6 (15.4)	4 (10)	1 (1.9)	7 (18.4)
No	33 (84.6)	36 (90)	53 (98.1)	31 (81.6)
The kinds of side effects experienced				
Drowsiness	3 (7.7)	1 (2.5)	–	5 (13.2)
Nausea and vomiting	1 (2.6)	3 (7.5)	1 (1.9)	2 (5.4)
Stomach crampy	1 (2.6)	1 (2.5)	–	1 (2.6)
Diarrhea	–	1 (2.5)	–	1 (2.6)
Itchiness and rash	–	3 (7.5)	–	1 (2.6)
Stiffness	–	1 (2.5)	–	1 (2.6)
Increasing heartbeat	1 (2.6)	–	–	–

(continued)

Table 6.2 (continued)

Variable	Number (%)			
	Markisa	Cimone innovation	Cikini Kramat	Gedong Pompa
Headache	–	1 (2.5)	–	1 (2.6)
Source of information about traditional medicine				
Parents/family	23 (59)	19 (47.5)	20 (37)	19 (50)
Paramedics' recommendation	3 (7.7)	3 (7.5)	1 (1.9)	–
Friends	10 (25.6)	14 (35)	17 (31.5)	13 (34.2)
Personal experience from previous illness	8 (17)	5 (12.5)	10 (18.5)	5 (13.2)
Local culture	8 (17)	9 (22.5)	6 (11.1)	7 (18.4)
Advertisement	–	1 (2.5)	5 (9.3)	2 (5.4)
Internet	3 (7.7)	3 (7.5)	6 (11.1)	2 (5.4)
Others	4 (102)	2 (5)	4 (7.4)	2 (5.4)
Books	3 (77)	–	–	–
Traditional medicine vendors	1 (2.6)	–	1 (1.9)	–
Homegrown herbal plants	32 (82)	31 (77.5)	8 (14.8)	2 (5.4)
Bangle	–	2 (5)	–	–
Binahong	4 (10.2)	4 (10)	3 (5.5)	–
Daun afrika	1 (2.6)	–	–	–
Daun dewa	7 (17.9)	–	–	–
Daun mangkuk	–	1 (2.5)	–	–
Daun katuk	2 (5.1)	–	1 (1.9)	–
Daun samsit	10 (25.6)	–	–	–
Daun sirih	2 (5.1)	5 (12.5)	1 (1.9)	–
Daun saga	1 (2.6)	1 (2.5)	–	–
Daun salam	1 (2.6)	–	–	–
Daun Hiang	1 (2.6)	–	–	–
Gingseng	4 (10.3)	–	1 (1.9)	–
Jarak	1 (2.6)	–	–	–
Jahe	9 (23)	19 (47.5)	1 (1.9)	–
Jahe merah	2 (5.1)	2 (5)	–	–
Jeruk nipis	–	1 (2.5)	–	–
Kencur	5 (12.8)	6 (15)	–	–
Kumis Kucing	3 (7.7)	4 (10)	–	1 (2.6)
Kunyit	14 (35.9)	21 (52.5)	1 (1.9)	1 (2.6)
Lempuyang	–	1 (2.5)	–	–

(continued)

Table 6.2 (continued)

Variable	Number (%)			
	Markisa	Cimone innovation	Cikini Kramat	Gedong Pompa
Lengkuas	6 (15.4)	2 (5)	–	–
Lidah buaya	1 (2.6)	–	1 (1.9)	–
Mahkota dewa	1 (2.6)	–	–	–
Pandan	–	1 (2.5)	–	–
Pecah beling	3 (7.7)	1 (2.5)	–	–
Pepaya	2 (5.1)	–	–	–
Sambung nyawa	2 (5.1)	1 (2.5)	1 (1.9)	–
Sambiloto	1 (2.6)	1 (2.5)	–	–
Sirsak	1 (2.6)	1 (2.5)	–	–
Sereh	2 (5.1)	2 (5)	–	–
Teleng	–	1 (2.5)	–	–
Temulawak	–	2 (5)	1 (1.9)	–
Medicine used during illness				
Traditional medicine	7 (17.9)	9 (22.5)	13 (24.1)	6 (15.8)
Modern medicine	16 (41)	19 (47.5)	20 (37)	21 (55.3)
Both	16 (41)	12 (30)	21 (38.9)	11 (28.9)

Table 6.3 Attitude toward traditional and herbal medicines

Variable	Number (%)			
	Markisa	Inovasi cimone	Cikini Kramat	Gedong Pompa
Some illnesses cannot be cured by modern medicines				
Agree	23 (48.9)	30 (54.5)	22 (31.4)	28 (43.8)
Undecided	8 (17)	9 (16.4)	10 (14.3)	9 (14.1)
Disagree	16 (34)	16 (29.1)	38 (54.3)	27 (42.2)
Traditional medicine is more effective than modern medicine				
Agree	16 (34)	17 (30.9)	16 (22.9)	13 (20.3)
Undecided	13 (27.7)	13 (23.6)	9 (12.9)	11 (17.2)
Disagree	18 (38.3)	25 (45.5)	45 (64.3)	40 (62.5)
Traditional medicine is safer than modern medicine				
Agree	32 (68.1)	32 (58.2)	35 (50)	23 (35.9)
Undecided	3 (6.4)	5 (9.1)	6 (8.6)	3 (4.7)
Disagree	12 (25.5)	18 (32.7)	29 (41.4)	38 (59.4)

Table 6.4 Knowledge of traditional and herbal medicines

Variable	Number (%)			
	Markisa	Inovasi Cimone	Cikini Kramat	Gedong Pompa
Traditional medicine does not have any side effect				
Yes	29 (61.7)	36 (65.5)	45 (64.3)	32 (50)
No	18 (38.3)	19 (34.5)	25 (35.7)	32 (50)
The use of traditional medicine needs adequate knowledge about the risk and benefit				
Yes	37 (78.7)	48 (87.3)	65 (92.9)	56 (87.5)
No	10 (21.3)	7 (12.7)	5 (7.1)	8 (12.5)

plants that were not intentionally planted by the respondents but could easily be found growing in their yards, such as *Kumis Kucing* and *Ciplukan.*

6.4 Discussion

Data shows that the percentage of traditional medicine use in four villages is quite high. A fundamental concept found here is the balance between mind and body, between different dimensions of individual bodily functioning and need, between individual and community, individual/community and environment, and individual and the universe (Bodeger, Kronenberg, and Burford 2007, p. 11). That is why, understanding the practice and use of traditional medicine in this research should not be done purely on a rigid scientific basis. The use of traditional medicine in the four urban villages shows a large percentage. More than 50% have used traditional medicine as a treatment preference (Kampung Markisa 39 respondents (83%), Kampung Cimone 40 respondents (72.7%), Kampung Cikini 54 respondents (77.1%), Kampung Gedong Pompa 38 respondents (59.4%)). More than 50% also used traditional medicine more than three times in the past year; Kampung Markisa 28 respondents (71.8%), Kampung Cimone 29 respondents (72.5%), Kampung Cikini 36 respondents (66.7%), and Kampung Gedong Pompa 28 respondents (73.7%).

The respondents were women (202 people 85.6%), housewives' jobs (170 people 72%), and the education level mostly junior high school/high school graduates. Previous research conducted in Kedah Malaysia by Maung Maung et al. (2019) also mentioned more women housewives who were participants in the FGD. Here the role of women as decision-makers in the family when there is a health issue seems to be prominent.

Evidence of the use of traditional medicine as local wisdom local wisdom whose knowledge has been obtained for generations can be seen from the high number of respondents who answer that parents/families are a source of information about traditional medicine; Kampung Markisa 23 respondents (59%), Kampung Cimone 19 respondents (47.5%), Kampung Cikini 20 respondents (37%), Kampung

Table 6.5 Types of herbal medicines consumed by local inhabitants

Name of herbs	Latin name	Markisa	Cimone	Cikini	G. Pompa	Total
Kunyit (Turmeric)	*Curcuma domestica* Vall or *Curcuma longa* Linn	16	13	18	15	62
Sirih Leaf	*Piper betle*	4	7	6	7	24
Jahe (Ginger)	*Zingiber officinale*	7	9	6	2	24
Asam (Tamarind)	*Tamarindus indica*	5	2	10	5	22
Kencur (Lesser Galangal)	*Kaemperia galanga* L	6	1	9	2	18
Samsit Leaf	*Gynura divaricata*	14	0	0	0	14
Binahong	*Anredera cordifolia*	4	3	5	0	12
Salam (Bay Leaf)	*Syzygium polyanthum*	1	2	0	5	8
Kumis Kucing	*Orthosiphon aristatus*	3	2	1	1	7
Sereh (Lemon Grass)	*Cymbopogon citratus*	1	1	2	3	7
Temulawak	*Curcuma zanthorrhiza*	2	0	4	1	7
Sambung Nyawa	*Gynura procumbens*	2	1	1	0	4
Ubi Samsit	*Gynura divaricata*	1	0	0	3	4
Ciplukan Leaf	*Physalis angulata* L	0	0	1	2	3
Daun Saga (Sage)	*Abrus precatorius*	3	0	0	0	3
Daun Sirsak (Soursop Leaf)	*Annona muricata*	0	2	0	1	3
Kulit Manggis (Mangosteen Skin)	*Garcinia mangostana* L.	0	1	2	0	3
Madu (Honey)	–	1	0	1	1	3
Pecah Beling	*Strobilanthes crispa*	2	1	0	0	3
Sambiloto	*Andrographis paniculata*	1	1	0	1	3
Seledri (Celery)	*Apium graveolens L*	0	1	1	1	3

Gedong Pompa 19 respondents (50%); a very large percentage of the use of traditional medicine in four urban villages. In this respect, health knowledge covered an appreciation of both the material and non-material properties of the herbal plants, as Bodeker and Burford suggested (2007: 3). More than 50% have used traditional medicine as a treatment preference; Kampung Markisa 39 respondents (83%), Kampung Cimone 40 respondents (72.7%), Kampung Cikini 54 respondents (77.1%), Kampung Gedong Pompa 38 respondents (59.4%). More than 50% also used traditional medicine more than three times in the past year; Kampung Markisa 28 respondents (71.8%), Kampung Cimone 29 respondents (72.5%), Kampung Cikini 36 respondents (66.7%), Kampung Gedong Pompa 28 respondents (73.7%).

The provisional conclusion is that the closer a village is to the site of modernity, the lesser the trust in the effectiveness of traditional medicine. It can be seen from

Kampung Cikini and Gedong Pompa which are in ring 1 of modernization (located in the city of Jakarta). In answering the question of whether there are diseases that cannot be cured by modern medicine, as many as 38 people (54.3%) in Kampung Cikini believe that there are no diseases that cannot be cured by modern medicine. The same is found in Kampung Gedong Pompa. The number of those who answered was absent and there was quite a balance, namely 27 people (42.2%) and 28 people (43.8%). Compared to Kampung Markisa and Kampung Cimone, only 34% and 29.1% believe that there are no diseases that cannot be cured by modern medicine.

There are various reasons for people to opt for herbal medicine, and the one frequently mentioned is that they were able to maintain a "sense of control" of their own therapies, which helps explain why sometimes people turned to herbal medicine for incurable diseases, such as diabetes, cancer, arthritis, or AIDS. They believed that conventional medicine had failed them (Pal and Shukla 2003: 282). Whereas, when people use herbal therapies for non-life-threatening illnesses, such as cold, sore throat, or bee sting, it is often because modern medicine is not immediately available, too inconvenient, costly or time-consuming (Winslow and Kroll 1998 qtd in Pal and Shukla 2003).

As many as 48.9% in Kampung Markisa and 54.5% in Kampung Cimone believe that there are diseases that cannot be cured by modern medicine. The above conclusions can also be validated regarding the question of whether traditional medicine is more effective than modern medicine. As many as 64.3% of respondents in Kampung Cikini and 62.5% of respondents in Kampung Gedong Pompa said that traditional medicine is no more effective than modern medicine. Various research has warned that traditional medicines do not easily lend themselves to the testing used for modern drugs primarily because many of the mixtures of compounds of these medicines are still unknown, and their compositions act together in a complex manner and, hence, the difficulty of scientific evaluation on them, especially in light of the absence of animal-testing (De Silva 2009, p. 64).

It seems that the negotiations between modernity versus locality or modern medicine versus traditional medicine are more apparent in urban villages that are far from ring 1 of modernization. The percentage of the use of traditional medicine is still large in the area of Kampung Cikini and Kampung Gedong Pompa, but many people are starting to doubt its effectiveness. Kampung Cikini and Kampung Gedong Pompa, which are close to modernization, seem to already have greater confidence in modern medicine, judging from the large number of respondents who agree that modern medicine is more effective than traditional medicine and the belief that there are no diseases that cannot be cured by modern medicine. Common health problems frequently encountered by rural communities, such as cold, cough, fever, and diarrhea, were also experienced by the kampung communities in urban and peri-urban Jakarta, and can be readily addressed through traditional solutions, as pointed out by Hariramamurthi et al. (2007, p. 168).

This is in contrast to Kampung Markisa and Cimone which are quite far away from modernization (in this case the city of Jakarta). The people in the two kampungs still use traditional medicine. There is a more balanced proportion in the population between people who believe and do not believe in the effectiveness of traditional medicine. On the edge of this modernization, the tug-of-war between the belief in modern medicine and traditional medicine is still quite strong. Respondents who believe in the effectiveness of traditional medicine are quite balanced with respondents who believe in the effectiveness of modern medicine (Kampung Markisa 34% and Kampung Cimone 30% believe in traditional medicine, Kampung Markisa 38.3% and Kampung Cimone 45.5% believe in modern medicine). However, when sick, quite a lot of respondents chose to use modern medicine; Kampung Markisa 41% and Kampung Cimone 47.5%. This shows the continued strength of negotiations between modernity versus locality in these regions.

Hoppenbrouwers, Sandarupa, and Donzelli wrote that, while modern medicine played a role in contemporary Toraja society, their research site, ancestral beliefs still guide health practices of the local people. Toraja is an exemplary case of rural society that is remotely located from any major urban setting. There was a dynamic interaction between medical discourse, common sense practices, and ancestral proscriptions (2017). Most information about the local inhabitants' use of traditional medicine was not available at the local office of the Public Health Department or remained unreported. They preferred treating illnesses using herbal medicine to taking the patients to hospital for treatment (2017, pp. 663–664). However, the inhabitants of the four kampungs of our research sites were showed a more welcoming attitude to modern medicine, and the resilience of the traditional use of herbal medicine can adequately be explained from the practical and pragmatic perspectives rather than a spiritual one, as clearly shown by the data resulted from the field work above.

The availability of land, which shrinks as the kampung is close to the city, also affects their use of traditional medicine—which is usually planted in the kampung area. In Kampung Markisa, 82% of respondents grow their own traditional medicines, in Kampung Cimone 77.5%, in Kampung Cikini 14.8%, and in Kampung Gedong Pompa 5.4%. The kampung dweller's folk wisdom about traditional medicine or knowledge about plants which can be used to relieve some ailments shows some relevance in their choice of using traditional medicine. Both Kampung Cikini and Gedong Pompa experienced a lack of adequate land, clean water, and sanitation. WHO and UNICEF estimated in 2000 that more than one billion people lack access to safe drinking water and that the global burden of diseases associated with poor water supply and sanitation equaled two billion cases of diarrhea and an annual death toll of 2.2 billion (WHO/UNICEF 2000, qtd in Hariramamurthi et al. 2007, p. 168).

In general, however, these findings confirm what Mueller and Mechler stated regarding the patterns of use of traditional and herbal medicine, namely that families use medicinal plants for curing illnesses, with a certain degree of effectiveness, because it is a form of home remedies that were applied by family members themselves and, therefore, they are free of any cost. In addition, information about these plants circulated freely between families belonging to the same kampung or community, and so they were not kept as secret (2005, p. 1). While modernization of traditional and herbal medicines seems to be preferred in order to close the gap between traditional medicine and modern medicine, which may be ultimately beneficial for rural and kampung communities that still rely directly on traditional medicine, modernization should not simply mean "westernization", as Liu argued (2011, p. 3).

So far, only a relatively small number of herbal medicines have been well studied, which include Echinacea, ginkgo, ginseng, and licorice. While many people in the above-mentioned kampungs believed that herbal products are safe in comparison with modern drugs, toxicological studies have shown that this is not always true. Basically, the safety of herbal plants used for medicinal purposes depends on dosage and period of use. The purification of some herbal extracts may increase their toxicity because, as the active components are concentrated, the concentration of toxic compounds may also be increased. Sometimes, the active components themselves are already toxic; thus, while the therapeutic effect is enhanced, the toxicity is also increased (Liu 2011, pp. 6–7).

6.5 Conclusion

Findings from the four kampungs show the urban kampung dwellers' not only familiarity with various herbs and plants to be used in traditional medicine but also their regular consumption of them. At the same time, they are also consuming modern medicine on the counter or prescribed medicine. The findings show that the four kampungs have different perceptions and considerations in choosing between the traditional and the modern medicine. The further away the kampung is from the urban centers, the more they are inclined to traditional medicine. One of the reasons can be found in the shrinking of land in the urban kampungs, and the affinity to modern lifestyle. Another context that relates to the large percentage of the use of traditional medicine in the kampungs is its accessibility, as it does not need a prescription and is relatively cheap. The research also finds indications, which still need further studies, of the role of women farmers' groups (KWT, *Kelompok Wanita Tani*) in growing the plants used for traditional medicine, and their roles in decision making about the use of medicine in their households.

Acknowledgements This study was supported by Direktorat Riset dan Pengabdian Kepada Masyarakat Universitas Indonesia (DRPM UI) under PINTERMIDI UI Research Grant No. 945/SK/R/UI/2019.

References

Bodeker, G., and G. Burford. 2007. Introduction. In *Traditional, complementary and alternative medicine: Policy and public health perspectives*, ed. G. Bodeker and G. Burford, 3–8. Imperial College Press.

Bodeker, G., F. Kronenberg, and G. Burford. 2007. Policy and public health perspectives on traditional, complementary and alternative medicine: An overview. In *Traditional, complementary, and alternative medicine: Policy and public health perspectives*, ed. Bodeker and Burford, 9–38. London: Imperial College Press.

De Silva, T. 2009. Traditional medicine in the developing and developed countries and expected trends in future. In *Traditional and alternative medicine: Research and policy perspectives*, ed. T. De Silva, T. Bahorun, M. Sahu, and M. H. Le, 57–66. London: Daya Publishing House.

Elfahmi, H. J. Woerdenbag, and O. Kayser. 2014. *Jamu*: Indonesian traditional herbal medicine toward rational phytopharmacological use. *Journal of Herbal Medicine* 4: 51–73. https://doi.org/10.1016/j.hermed.2014.01.002.

Hariramamurthi, G., P. Venkatasubramanian, P. M. Unnikrishnan, and D. Shankar. 2007. Home herbal gardens—a novel health security strategy based on local knowledge and resources. In *Traditional, complementary and alternative medicine: Policy and public health perspectives*, ed. Bodeker and Burford, 167–183. London: Imperial College Press.

Hoppenbrouwers, T., S. Sandarupa, and A. Donzelli. 2017. From the womb to the tree: Child-rearing practices and beliefs among the Toraja of Sulawesi. *Wacana Journal of the Humanities of Indonesia* 18 (3): 658–691. https://doi.org/10.17510/wacana.v18i3.632.

Kantor Menteri Negara Kependudukan dan Lingkungan Hidup. 1992. *Indonesian Country Study on Biological Diversity*. Kantor Menteri Negara Kependudukan dan Lingkungan Hidup.

Kautsar, A. P., M. F. S. Ayunovani, and E. Surahman. 2016. The influence of demographic, social system, communication system, and herbal characteristics on purchase decisions of herbal medicine in Indonesia. *Journal of Economics, Business and Management* 4 (3): 235–238. https://doi.org/10.7763/JOEBM.2016.V4.396.

Liu, W. J. H. 2011. Introduction to traditional herbal medicines and their study. In *Traditional herbal medicine research methods: Identification, analysis, bioassay, and pharmaceutical and clinical studies*, ed. W. J. H. Liu, 1–26. London: Wiley.

Lubis, A. M. 2009. Indonesia country report on herbal medical plants and traditional herb remedies. In *Traditional and alternative medicine: Research and policy perspectives*, ed. T. De Silva, T. Bahorun, M. Sahu, and M. H. Le, 532–539. London: Daya Publishing House.

Maung, T. M., S. G. Deborah, and A. A. Tun. 2019. Traditional medicine vs. modern medicine in the rural area of Kedah State, Malaysia. *Journal of Pharmacy and Biological Sciences (IOSR-JPBS)* 14 (1): 05–08. https://doi.org/10.9790/3008-1401050508.

Mueller, M. S., and E. Mechler. 2005. *Medicinal plants in tropical countries: Traditional use—Experience—Facts*. Stuttgart: New York Thieme.

Nurhayati, and L. Widowati. 2017. The use of traditional health care among Indonesian family. *Health Science Journal of Indonesia* 8 (1): 30–35. https://doi.org/10.22435/hsji.v8i1.5600.30-35.

Pengpid, S., and K. Peltzer. 2018. Utilisation of traditional and complementary medicine in Indonesia: Results of a national survey in 2014–15. *Complementary Therapies in Clinical Practices* 33: 156–163. https://doi.org/10.1016/j.ctcp.2018.10.006.

Pal, S. K., and Y. Shukla. 2003. Herbal medicine: Current status and the future. *Asian Pacific Journal of Cancer Prevention* 4 (4): 281–288.

Zakaria, F., & Zainal, H. 2017. Traditional Malay medicine in Singapore. *Indonesia and the Malay World* 45 (131): 127–144. https://doi.org/10.1080/13639811.2017.1275156.

Chapter 7
Misuse of Mixed Prescribed Medication (*Obat Rombongan*) as Self-Medication: Lesson Learned from Gedong Pompa Urban Kampung, Jakarta, Indonesia

Bantari W. K. Wardhani, Indira Rahmawati, Juwita Sari, Nabila, and Melva Louisa

Abstract In Jakarta, urban kampung provides a reasonable informal settlement for low-income people and newcomers in big cities. Hence, it consists of a high diversity of community characteristics. Recently, a fast information exchange via social and electronic media has affected the behaviors of society. Behavior toward self-medication is also affected by the social media, education, and location of settlements. One of the urban kampung located in North Jakarta, Gedong Pompa, was known for its resilience and socio-spatial adaptation to internal and external pressure due to its community's limitations. However, how they behaved regarding self-medication was not known. The present study described self-medication behavior in Kampung Gedong Pompa. In-depth interviews have brought us to misuse of a mixture of prescribed medicines (the people call it *obat rombongan*). The mix of

B. W. K. Wardhani
Doctoral Program in Biomedical Sciences, Faculty of Medicine, Universitas Indonesia, Jakarta Pusat, Indonesia

Research Center for Pharmaceutical Ingredients and Traditional Medicine, BRIN, National Research and Innovation Agency, Jakarta Pusat, Indonesia

B. W. K. Wardhani
e-mail: bantariwisynu@gmail.com

I. Rahmawati · J. Sari · Nabila
Faculty of Medicine, Universitas Indonesia, Jakarta, Indonesia
e-mail: indirarara10@gmail.com

J. Sari
e-mail: juwita.sari@ui.ac.id

Nabila
e-mail: nabilamulyawan7@gmail.com

M. Louisa (✉)
Department of Pharmacology and Therapeutics, Faculty of Medicine, Universitas Indonesia, Jakarta Pusat, Indonesia
e-mail: melva.louisa@gmail.com

M. Budianta et al. (eds.), *A Liveable Kampung*, Engaging Indonesia,
https://doi.org/10.1007/978-981-96-7866-2_7

prescribed medications that people can buy in nearby stores contains antibiotics, corticosteroids, NSAIDs, and vitamins. However, we found out that they had a good perception regarding self-medication. Instead of increasing their access to cheaper healthy expenditure, their inappropriate behavior might lead to antibiotic resistance and several drug-related problems.

Keywords Misuse drug · Self-medication · Obat rombongan · Urban household · Urban kampung

7.1 Introduction

People in many communities have long practiced self-medication (SM). Recently, the trends have increased, especially during the COVID-19 pandemic in 2020 (Onchonga 2020). Self-medication (SM) is a practice to use drugs by an individual without a prescription or a doctor's diagnosis and supervision from health workers (DepKes RI 2006; Onchonga 2020). SM is used to alleviate mild sickness symptoms quickly and rapidly, reducing the burden on healthcare facilities, particularly in understaffed or inaccessible distant places (Onchonga 2020). Self-medication is carried out by people around the world with different prevalence in each country. The prevalence of SM in the US is 14%, in China 45.4%, and in Indonesia, it is 62.7% (Bennadi 2014; Lei et al. 2018). In several population subgroups, the prevalence of SM also varies, 98.9% and 57.9% in Iranian and Indian women, respectively. (Karimy et al. 2019; Subashini et al. 2017). In males, the prevalence of SM practice was 79.8% in India (Subashini et al. 2017). In addition, the prevalence of SM in geriatrics also varies. In China, it was reported at 35.2% (Xu et al. 2020a, b), while in Indonesia, it was 94.1% (Dwi Tama et al. 2020).

Together with the Ministry of Health of the Republic of Indonesia, the World Health Organization has long launched SM to expand public access to drugs (Permenkes 1993; WHO 1998). SM can help provide rapid, convenient access at a reasonable cost of medicines without going to health facilities first (Lei et al. 2018; Onchonga 2020). However, studies have shown the deleterious impact of the irrational practice of SM. Irrational implementation of SM is harmful due to the possibilities of inappropriate use of drugs concerning their type, duration, and indications (Lei et al. 2018). SM is influenced by the sources of information received by the people on media, experiences of family/neighbors/colleagues, education, and its location (Bennadi 2014; Limaye et al. 2018). A prior study by Aqeel in Pakistan has demonstrated the association between regional location and SM (Aqeel 2014). Two studies of SM in India have also confirmed the association between urban areas and the community's pattern of SM (Cecyli and Pragathi 2020; Limaye et al. 2018).

Urban kampungs, sometimes referred to as slum regions and densely inhabited districts located in and around big cities, had shown a particular type of practice regarding health-related practice. Referring to the study conducted by Putri and Herlily (2020a, b), an urban village is an informal residential area in the middle of the city with a community that is vulnerable to internal and external pressures and the ability to adapt at the same time. Internal and external pressures include limited fulfillment of daily needs, including housing, sanitation, and access to clean water. However, these limitations allow urban villages to adapt based on community (Putri and Herlily 2020a). Shirleyana et al. (2018), in their study, revealed that urban villages in the middle of urban areas provide affordable housing for low-income people and migrants in big cities (Shirleyana et al. 2018). Many previous studies have demonstrated resilience and socio-spatial adaptation (Putri and Herlily 2020a; Samosir et al. 2020; Shirleyana et al. 2018). Nevertheless, there has been no study describing the SM perception of the people living in urban kampung areas in Jakarta areas.

Kampung Gedong Pompa is located in the Penjaringan area, North Jakarta. The population density from 2013 to 2018 has increased dramatically, in line with the economic growth in the area. The fast population growth is influenced by the village's location close to industrial areas, markets, and the beach. In addition, the price of rented houses in this area is relatively affordable (Putri et al. 2020b). Therefore, this village is suitable for the needs of migrants without formal skills to work informally in the surrounding industrial areas and low-income communities. The resilience of urban kampung in adapting to changes in internal and external pressures includes various aspects, including health and well-being perspectives (Putri and Herlily 2020b). With such a population character, rational SM can be a beneficial solution to fulfill the need for affordable and readily available drugs. However, the implementation of irrational SM can cause antibiotic resistance, unnecessary drug use, and an increased risk of drug toxicity (Karimy et al. 2019). Therefore, it is necessary to describe the perception and US of SM on households in urban kampung residents, particularly Gedong Pompa residents, Penjaringan, North Jakarta.

7.2 Methods

7.2.1 Study Design

This research applied cross-sectional descriptive study in Gedong Pompa, Penjaringan, North Jakarta, during February 2020. The ethical approval of the current study was approved by the Ethics Committee Faculty of Medicine Universitas Indonesia with the Number of KET-1060/UN2.F1/ETIK/PPM.00.02/2019.

7.2.2 Study Participants

A simple random sampling was used to enroll 150 households' respondents. Each respondent gave informed consent after receiving clear information regarding the

Table 7.1 List of questions included in the questionnaire

Questions	Answer	
Please state if you agree with the following statements		
1. Over-the-counter (OTC) drugs used for self-medication are safe	Y	N
2. Over-the-counter (OTC) drugs are economical and easy to obtain	Y	N
3. The quality of over-the-counter (OTC) drugs are not affected by storage conditions, temperature, humidity, and light		
4. I do not prefer to use OTC drugs when the member of my family or I are sick		
5. OTC drugs are safe. However, it is better to ask a Health Care Provider for advice on using them		

aim of the study and the procedures. Inclusion criteria were men or women aged above 18 years old, permanent residents of Kampung Gedong Pompa, and providing signed informed consent.

7.2.3 Data Collection

We collected demography data and a pre-validated questionnaire to record the practice and perspectives of SM in the residents of Kampung Gedong Pompa. The questions were given in Bahasa Indonesia, and the translation is provided in the following table (Table 7.1).

7.2.4 Data Analysis

The demographics of subjects were shown in the frequency table. SM perceptions of the people in Gedong Pompa were provided qualitatively.

7.3 Results

This cross-sectional study recruited 150 respondents in Kampung Gedong Pompa, North Jakarta.

7.3.1 Respondents Demography

Respondents in this study were between 19 and 92 years old, with most of the proportion being women. The respondents' education levels were mainly high school graduates or the equivalent. At the same time, most of the primary income earners in the family are casual laborers with daily income as farmers, ranchmen, fishers, and miners (Table 7.2).

Table 7.2 Demographic characteristics

Variable	Frequency (%)
Sex	
Male	37 (24.7)
Female	113 (75.3)
Age in years	
20–39	62 (41.3)
40–59	71 (47.3)
≥ 60	17 (11.3)
Academic status	
No educational background	44 (29.3)
Basic elementary education (Primary and Junior High School)	75 (50.0)
High school and University	31 (20.7)
Occupation	
Fishermen	10 (6.7)
Entrepreneur	24 (16.0)
Government or private sector employee	39 (26.0)
Casual laborers	77 (51.3)
House ownership	
Owned	79 (52.7)
Rent	46 (30.7)
Office residence	2 (1.3)
Rent-free	22 (15.3)
Toddler or elderly in the household	
Toddler(s)	53 (35.3)
Elderly	21 (14.0)
Toddler(s) and elderly	6 (4.0)
No toddler or elderly	76 (50.7)

7.3.2 Perceptions Toward Self-Medication

The majority of respondents (90%) living in the urban settlement of Gedong Pompa agreed that SM is a more economical and easier way to obtain treatment rather than going to a health facility for an examination and prescription from a health care provider. Most respondents (60%) agreed to use SM when they got ill.

SM can cause drug use errors due to lack of education regarding the safety of drug use and storage and supervision and recommendation of health workers, including doctors, pharmacists, nurses, and midwives. In the present study, about half of the respondents agreed that the drugs used in SM were safe (47%). Regarding drug storage, 59% of respondents agreed to pay attention to storage conditions, temperature, humidity, and light. Furthermore, 88% of respondents agreed to seek a doctor's advice before using drugs in SM.

7.3.3 Inappropriate Use of Drugs in SM

Over-the-counter drugs and limited over-the-counter drugs are classes of drugs considered relatively safe to use in SM (Ministry of Health 2006). The present study found that there were practices of inappropriate use of medicines in SM. The practice includes prescription drugs with no supervision of health care providers, which were found in 18.7% of respondents. The type of medicines used were drugs for complaints of aches. Stall owners around the area mixed several drugs to overcome complaints about headaches, stomach aches, and toothaches. The mixed medicines consisted of at least 4 of the following drugs: antibiotics (tetracyclines), analgesics (mefenamic acid), corticosteroids (dexamethasone or prednisone), antihistamines (chlorpheniramine maleate), antidiarrheals (loperamide), and B-complex vitamins.

7.4 Discussion

Self-medication has varying prevalence in different areas, urban, suburban, or rural areas. Factors affecting SM vary in age, income, education level, family history, access to information about drugs, and drug availability (Bennadi 2014; Lei et al. 2018; Limaye et al. 2018). Access to drugs and advertisements is easily obtained through the internet, electronic media, and social media. In addition, the world is currently experiencing a pandemic condition that limits the mobilization of people. It also increases the trend of self-medication in the community (Malik et al. 2020; Onchonga 2020).

The high prevalence of SM in various households, if misused, without being accompanied by good perceptions, attitudes, and knowledge is prone to drug-related problems in the use of drugs. Self-medication is the most significant cause of antibiotic resistance due to inappropriate use and without the supervision of health workers (Rehman et al. 2020). OTC drugs are relatively safe to use in SM (Bennadi 2014; WHO 2000). In Indonesia, through the Ministry of Health, the government has limited over-the-counter and over-the-counter drugs for SM (Brata et al. 2016; Ministry of Health 2006; Regulation of the Minister of Health 1993). However, non-OTC drugs are often used in SM, such as antibiotics, antivirals, and corticosteroids (Cecyli and Pragathi 2020; Lei et al. 2018; Quispe-Cañari et al. 2021).

This study was done to investigate the perception of the urban settlements, Kampung Gedong Pompa, Penjaringan, North Jakarta on self-medication. Perception of SM affects safe and rational drug use behavior (Masud et al. 2020). However, a good perception of respondents is not always followed by responsible SM (Seam et al. 2018; Susheela et al. 2018).

This study explores household respondents' perceptions of the safety, accessibility, and ease of access and storage conditions for drugs used for SM. In addition, the tendency to choose SM and the need for advice from health workers were also investigated.

The practice of SM is considered economical and easier to access by the community than going to a health facility seeking treatment for symptoms or complaints experienced. Our findings align with the goals of SM set by the WHO to increase public access to drugs and reduce the burden on health facilities (Malik et al. 2020; Guidelines for the Regulatory Assessment of Medicinal Products for Use in Self-Medication 2000). In addition, most respondents in this study agreed that SM provides more accessible and more economical access to drugs. The present finding is supported by other studies in India, Ethiopia, China, and Iran. Susheela et al. (2018) reported that 26.8% of student respondents majoring in health in India chose SM because it saves costs and time for health facilities (Susheela et al. 2018). The perception of cheaper SM was also reported by 21.5% of the household population in Ethiopia and 15.1% in China (Jember et al. 2019; Xu et al. 2020a, b). The high proportion in this study could be due to 51.3% of respondents working as casual laborers. In this job, the income earned is daily. In SM practice, respondents admitted that they did not need to go to a health facility and queue to consult a doctor or other health worker to get a prescription. Day laborers are usually also paid according to the hours worked. In addition, the people in the urban village have lower-middle-income (Putri and Herlily 2020a). The findings were illustrated in Table 7.1, which showed that only half of the respondents have ownership of their current residence. Therefore, the respondents avoided ways of obtaining drugs that cost time and money.

Regarding perceptions on SM, the majority showed an adequate perception of the storage of drugs used in SM. A total of 59% agreed to pay attention to storage conditions, exposure to light, temperature, and humidity in the storage of drugs used in SM. Perceptions of the safety of drugs used in SM are reasonably good. More than half of the respondents disagreed with the statement that the use of non-prescription drugs in SM was safe (53%). Perceptions of drugs used without a prescription are safe and are mainly influenced by the history of side effects experienced. Without sufficient knowledge about drug safety, some respondents have a perception that the drugs used are safe because they have never experienced side effects and always get satisfaction (complaints are cured) from the treatment used in SM. However, the majority (88%) agreed that it is better to seek advice from doctors or other health workers (nurses, midwives, and pharmacists) if they use drugs in SM.

Another compelling finding in the present study was the use of mixed prescription drugs by the people living in Gedong Pompa with no prescription and supervision from health care providers, despite the satisfactory perception of SM practice. The people in Gedong Pompa called the mixed prescription drugs "*obat rombongan.*" Mixed prescription medicines were commonly obtained at the local stalls and treated headaches, stomachaches, diarrhea, and toothaches. The mixed medicines consisted of at least four medicines: antibiotics, analgesics, corticosteroids, antihistamines, antidiarrheals, and vitamins. Each drug class contained one or two tablets in one package. The antibiotic used was tetracycline, which the respondents commonly called "super tetra." The analgesic contained in the group of drugs was mefenamic acid. Corticosteroids contained in the drug group are dexamethasone and prednisone. Antihistamines included in the group of drugs are chlorpheniramine maleate. There is also a vitamin B complex that is added to the group's medication for all complaints. For diarrhea complaints, there was additional loperamide in the group's drug collection.

The use of prescription mixed drugs in the practice of SM in Kampung Gedong Pompa is considered inappropriate use of drugs not following the guidelines from WHO (2000) and the Regulation of the Minister of Health of the Republic of Indonesia (Permenkes 1993). Rodrigues (2020) and Aslaam (2020) also stated the same thing. The drugs used in group medicine are antibiotics and hard drugs that should be obtained using a doctor's prescription (Aslam et al. 2020; Rodrigues 2020). Using one of these drugs alone is included in using the wrong drug (misuse), let alone used together.

Antibiotics were taken by the people of Kampung Gedong Pompa when there were symptoms (*pro renata*), only once or twice in one period of use. They will stop the medications once the symptoms subside. The people have used antibiotics with inappropriate indications as well as regiments. Acute gastrointestinal disorders such as diarrhea generally do not require antibiotics (Riddle et al. 2016). Empirical use of antibiotics in diarrhea is carried out in conditions of diarrhea accompanied by fever, blood, duration of more than one week, or impaired immune system (Zollner-Schwetz and Krause 2015). Tetracycline, the antibiotic used at Kampung Gedong Pompa, is an antibiotic indicated for cholera, given for three days. Tetracyclines can be replaced by doxycycline, cotrimoxazole, norfloxacin, and ciprofloxacin (Cashburn-Jones and Farthing 2004). Recently, however, the drug of choice for diarrhea caused by Vibrio cholerae is doxycycline instead of tetracycline (Barr and Smith 2014). In addition to diarrhea, super tetra antibiotics are also used for other complaints, toothache, and aches. The use of these two complaints without a doctor's prescription is also an irrational SM practice. The inappropriate use of antibiotics can, of course, increase the risk of resistance (Rehman et al. 2020).

Loperamide is oral anti-diarrhea that acts as an antimotility. Loperamide slows the movement of intraluminal fluid and increases the reabsorption of fluid in the intestine. In addition to increasing motility, loperamide reduces mucosal secretions. Therefore, loperamide may exacerbate bloody diarrhea or intestinal inflammation (Riddle et al. 2016). Loperamide is indicated in moderate to severe diarrhea with fever symptoms for more than three days or less than two days without fever and without a history of travel outside the city or abroad (travelers' diarrhea) (Barr and Smith 2014). Indications of loperamide are also limited only for diarrhea not accompanied by blood (Riddle et al. 2016).

One of the mixed drug components was mefenamic acid, a non-steroidal anti-inflammatory drug (NSAIDs). The use of analgesics is frequent in self-medication, as is the case with studies conducted by Seam et al. (2018) and Patel et al. (2014) in India, Gogazeh (2020) in Jordan, and Quispe-Canari et al. (2021) in Peru. This can be related to the majority of respondents using self-medication for complaints of headaches, menstrual pain, toothache, and muscle pain (Ahmad et al. 2014; Gogazeh 2020; Quispe-Cañari et al. 2021; Seam et al. 2018). Using analgesics without proper education can lead to incorrect use, such as improper use (on an empty stomach). Excessive use of analgesics is also associated with the incidence of gastric mucosal injury (Fujino et al. 1985; Oktaviyanti et al. 2016). Analgesics in the NSAID group work by inhibiting the cyclooxygenase enzyme. Inhibition of this enzyme results in a decrease in prostaglandins. Prostaglandins are needed to suppress gastric acid secretion, stimulate mucus and bicarbonate secretion, and protect against substances that can damage gastric acid (Bjarnason et al. 2018).

Corticosteroids, including dexamethasone and prednisone, were also part of the cocktail drugs used. In self-medication, the use of corticosteroids, such as dexamethasone, also occurred in Peru (Quispe-Cañari et al. 2021) and India (Ahmad et al. 2014). Short-term use of oral corticosteroids in the general adult population in the US has been reported to be associated with an increased incidence of sepsis, venous thromboembolism, and fractures (Waljee et al. 2017).

The majority of household respondents in this study had vulnerable groups in drug use, toddlers, the elderly, or both. Elderly and toddlers are subgroups of populations that are more prone to adverse drug effects and may require dose adjustments according to doctors' judgment who prescribe drugs, especially a potent drug such as antibiotics (Xu et al. 2020a, b; Xu et al. 2020a, b). Inappropriate drug use behavior (misused or overused) from the head of the family might lead to inappropriate and unnecessary drug use in children. A study in China stated that 32% of Chinese parents used antibiotics for self-medication without consultation and examination with health workers (Xu et al. 2020a, b). Studies in Pakistan also show a similar trend. A total of 47% of parents in Pakistan choose to do self-medication for their children aged 1–5 years, with a proportion using antibiotics of 34% (Gohar et al. 2017). One of the limitations of this study was that it did not look further into whether drugs such as group drugs were used for the elderly and toddlers. However, the respondents who answered the interviews in this study were the decision-makers in the household. If the decision holder has the wrong behavior in self-medication, this can result in family members, including the elderly and toddlers.

The educational background of respondents could be a determinant in self-medication behavior in Gedong Pompa. A small proportion of decision-makers in households graduated from high school or vocational school. There is only one person who has graduated from a university at the undergraduate level. One-third of the respondents had not even graduated from school and could not read well. Educational background is one factor that determines the behavior of using SM (Lei et al. 2018; Rodrigues 2020; Selvaraj et al. 2014). As studied in China, the respondents' education level correlates with reading and understanding of drug use information (Lei et al. 2018). A study in Maputo (Rodrigues 2020) showed that educational background has a vital role in the autonomy of drug use by an individual. The study also revealed an educational "gap" between doctors and the patient community, leading to terminology misunderstandings for symptoms, drugs, and other medical information (Rodrigues 2020). Groups of respondents with higher levels of education had less use of self-medication than those with lower levels of education (Selvaraj et al. 2014).

The location of urban settlements in the middle of a big city is also considered in influencing self-medication behavior in Gedong Pompa, Penjaringan, North Jakarta. The urban settlement is one of the portraits of the community's adaptation to the modernization of big cities. The modernization that occurs follows the concept of compressed modernity. In this theory, cities in Asian countries are forced to follow the flow of globalization and are compressed into "modern." One of the dimensions that have changed is traditional behavior toward modern, practical, and easy (Chang 2017). Traditional behavior includes switching to modern medicine and using modern medicines that are easier and more practical without going through a processing process. Self-medication behavior is also an adaptation of urban village communities to get easier access to treatment than health facilities. Information about drugs can be obtained easily through the internet or the experience of the closest people, such as neighbors. As a result, inappropriate drug use behavior in one person develops into household behavior and then becomes wrong in the community.

Ideally, self-medication behavior must be followed by continuing counsel and education from health workers, including doctors, nurses, midwives, and pharmacists from the nearest health centers, both government-owned and private (medicine clinics and pharmacies). The empowerment of healthcare cadres and youth groups also need to be considered in correcting drug use errors. In addition, government supervision needs to be carried out on a micro-scale to urban villages to break the distribution chain of solid drugs that should not be used in self-medication.

The present study was done just prior to COVID-19 pandemic (February–March 2020). Therefore, this data may still be similar to the present conditions or more due to the pandemic scare. However, the inappropriateness of SM practice could still be in place.

In conclusion, the residents of Kampung Gedong Pompa have a proper perspective on self-medication. Nevertheless, inappropriate and unsafe self-medication practices persist, using cocktail prescription medications with no supervision from health care providers. The best way to change the inappropriate practice of prescribed medicine is by providing continuous educational intervention.

Acknowledgements This study was supported by Direktorat Riset dan Pengabdian Kepada Masyarakat Universitas Indonesia (DRPM UI) under PINTERMIDI UI Research Grant No. 945/ SK/R/UI/2019.

References

Ahmad, A., I. Patel, G. Mohanta, and R. Balkrishnan. 2014. Evaluation of self-medication practices in the rural area of town Sahaswan in Northern India. *Annals of Medical and Health Sciences Research* 4 (8): 73–78. https://doi.org/10.4103/2141-9248.138012.

Aqeel, Tahir, Arham Shabbir, Hina Basharat, Malika Bukhari, S. Mobin, H. Shahid, and S. A. Waqar. 2014. Prevalence of self-medication among urban dan rural population of Islamabad, Pakistan. *Tropical Journal of Pharmaceutical Research, 13*(4), 627–633.

Aslam, A., M. Gajdács, C. S. Zin, N. S. A. Rahman, S. I. Ahmed, M. Z. Zafar, and S. Jamshed. 2020. Evidence of the practice of self-medication with antibiotics among the lay public in low-and middle-income countries: A scoping review. *Antibiotics* 9 (9): 1–17. https://doi.org/10.3390/antibiotics9090597.

Barr, W., and A. Smith. 2014. Acute diarrhea in adults. *Australian Family Physician* 89 (3): 180–190. https://doi.org/10.1136/bmj.b1877.

Bennadi, D. 2014. Self-medication: A current challenge. *Journal of Basic and Clinical Pharmacy* 5 (1): 19–23. https://doi.org/10.4103/0976-0105.128253.

Bjarnason, I., C. Scarpignato, E. Holmgren, M. Olszewski, K. D. Rainsford, and A. Lanas. 2018. Mechanisms of damage to the gastrointestinal tract from non-steroidal anti-inflammatory drugs. *Gastroenterology* 154 (3): 500–514. https://doi.org/10.1053/j.gastro.2017.10.049.

Brata, C., C. Fisher, B. Marjadi, C. R. Schneider, and R. M. Clifford. 2016. Factors influencing the current practice of self-medication consultations in Eastern Indonesian community pharmacies: A qualitative study. *BMC Health Services Research* 16 (1): 1–10. https://doi.org/10.1186/s12913-016-1425-3.

Cashburn-Jones, A., and M. J. Farthing. 2004. Management of Infectious Diarrhoea. *Gut* 53 (2): 296–305. https://doi.org/10.1136/gut.2003.022103.

Cecyli, C., and T. Pragathi. 2020. Assessment of knowledge and practice of self-medication among the urban and rural population. *Drug Invention Today* 13 (7): 1059–1064.

Chang, K.-S. 2017. Compressed Modernity. *The Wiley-Blackwell Encyclopedia of Social Theory*. https://doi.org/10.1002/9781118430873.est0839.

Departemen Kesehatan, R. I. 2006. *Pedoman penggunaan obat bebas dan bebas terbatas.*epartemen Kesehatan Republik Indonesia (*Guidelines for the consumption of on the counter medicine and restricted medicine*, Department of Health, Republic of Indonesia). Departemen Kesehatan Republik Indonesia, pp. 10–79.

Dwi Tama, T., N. Herya Ulfa, and A. Hapsari. 2020. Who Spends More on Health Expenditures among the Elderly? Men or Women? *KnE Social Sciences* 4 (10): 154–160. https://doi.org/10.18502/kss.v4i10.7402.

Fujino, S., Y. Suzuki, and T. Tanaka. 1985. Cost-benefit analysis of medicinal treatment for gastric ulcers. A long-term model including healing and recurrence. *Health Policy* 5 (1): 45–72. https://doi.org/10.1016/0168-8510(85)90065-X.

Gogazeh, E. 2020. Dispensing errors and self-medication practice observed by community pharmacists in Jordan. *Saudi Pharmaceutical Journal* 28 (3): 233–237. https://doi.org/10.1016/j.jsps.2020.01.001.

Gohar, U. F., S. Khubaib, and A. Mehmood. 2017. Self-medication trends in children by their parents. *Journal of Developing Drugs*. https://doi.org/10.4172/2329-6631.1000173.

Guidelines for the Regulatory Assessment of Medicinal Products for Use in Self-Medication, 1 WHO/EDM/QSM/00.1 1. 2000. https://apps.who.int/iris/handle/10665/66154

Jember, E., A. Feleke, A. Debie, and G. Asrade. 2019. Self-medication practices and associated factors among households at Gondar town, Northwest Ethiopia: A cross-sectional study. *BMC Research Notes* 12 (1): 1–7. https://doi.org/10.1186/s13104-019-4195-2.

Karimy, M., M. Rezaee-Momtaz, M. Tavousi, A. Montazeri, and M. Araban. 2019. Risk factors associated with self-medication among women in Iran. *BMC Public Health* 19 (1): 1–7. https://doi.org/10.1186/s12889-019-7302-3.

Lei, X., H. Jiang, C. Liu, A. Ferrier, and J. Mugavin. 2018. Self-medication practice and associated factors among residents in Wuhan, China. *International Journal of Environmental Research and Public Health* 15 (1): 68. https://doi.org/10.3390/ijerph15010068.

Limaye, D., V. Limaye, G. Fortwengel, and G. Krause. 2018. Self-medication practices in urban and rural areas of western India: A cross-sectional study. *International Journal of Community Medicine and Public Health* 5 (7): 2672. https://doi.org/10.18203/2394-6040.ijcmph20182596.

Malik, M., M. J. Tahir, R. Jabbar, A. Ahmed, and R. Hussain. 2020. Self-medication during Covid-19 pandemic: Challenges and opportunities. *Drugs and Therapy Perspectives* 36 (12): 565–567. https://doi.org/10.1007/s40267-020-00785-z.

Masud, M., Z. A. Mohamed, N. F. Azman, and M. A. A. Rahim. 2020. The practice, perception, and awareness of self-medication for dental pain in Malaysian dental students. *Journal of International Dental and Medical Research 13* (2), 697–703. http://www.jidmr.com/journal/wp-content/uploads/2020/06/50-D20_1036_Mahyunah_Masud_Malaysia.pdf

Peraturan Menteri Kesehatan No. 919/MENKES/PER/X/1993, *Tentang kriteria obat yang dapat diserahkan tanpa resep*, 00226020 1 (1993).

Oktaviyanti, I. K., I. Thalib, and E. Suhartono. 2016. The protective efficacy of quercetin on mefenamic acid-induced gastric mucosal damage. *International Journal of Pharmaceutical and Clinical Research* 8 (10): 1390–1395.

Onchonga, D. 2020. A Google Trends study on the interest in self-medication during the 2019 novel coronavirus (COVID-19) disease pandemic. *Saudi Pharmaceutical Journal* 28 (7): 903–904. https://doi.org/10.1016/j.jsps.2020.06.007.

Putri, N. S., and Herlily. 2020a. Communal spatial adaptation in response to urban kampung vulnerability, study case: Public toilet of Kampung Muka, North Jakarta. *AIP Conference Proceedings*. https://doi.org/10.1063/5.0004812.

Putri, N. S., and Herlily. 2020b. Investigating community adaptability and resilience in Urban Kampung, study case: Water and sanitation infrastructure of Kampung Muka, North Jakarta. *IOP Conference Series: Earth and Environmental Science*. https://doi.org/10.1088/1755-1315/452/1/012095.

Putri, R. A. Siti Delima Amanda, and Chotib. 2020. Detecting informal settlement development in Penjaringan sub-district, North Jakarta, Indonesia. *IOP Conference Series: Earth and Environmental Science*. https://doi.org/10.1088/1755-1315/436/1/012008.

Quispe-Cañari, J. F., E. Fidel-Rosales, D. Manrique, J. Mascar6-Zan, K. M. Huamán-Castill6n, S. E. Chamorro-Espinoza, H. Garayar-Peceros, V. L. Ponce-L6pez, J. Sifuentes-Rosales, A. Alvarez-Risco, J. A. Yáñez, and C. R. Mejia. 2021. Self-medication practices during the COVID-19 pandemic among the adult population in Peru: A cross-sectional survey. *Saudi Pharmaceutical Journal* 29 (1): 1–11. https://doi.org/10.1016/j.jsps.2020.12.001.

Rehman, M., S. Ahmed, U. Ahmed, K. Tamanna, M. S. Sabir, and Z. Niaz. 2020. An overview of self-medication: a major cause of antibiotic resistance and a threat to global public health. *Journal of the Pakistan Medical Association* 11: 1–22. https://doi.org/10.47391/jpma.1331.

Riddle, M. S., H. L. Dupont, and B. A. Connor. 2016. ACG clinical guideline: Diagnosis, treatment, and prevention of acute diarrheal infections in adults. *American Journal of Gastroenterology* 111 (5): 602–622. https://doi.org/10.1038/ajg.2016.126.

Rodrigues, C. F. 2020. Self-medication with antibiotics in Maputo, Mozambique: Practices, rationales, and relationships. *Palgrave Communications* 6 (1): 1–12. https://doi.org/10.1057/s41599-019-0385-8.

Samosir, N. S., Herlily, and K. R. Kurniawan. 2020. Food production and distribution from home industry in Kampung Kota Jakarta: An exploration of urban spatial trajectory. *IOP Conference Series: Earth and Environmental Science*. https://doi.org/10.1088/1755-1315/452/1/012005.
Seam, M., R. Bhatta, B. Saha, A. Das, M. Hossain, S. Uddin, P. Karmakar, M. Choudhuri, and M. Sattar. 2018. Assessing the perceptions and practice of self-medication among Bangladeshi undergraduate pharmacy students. *Pharmacy* 6 (1): 6. https://doi.org/10.3390/pharmacy6010006.
Selvaraj, K., G. S. Kumar, and A. Ramalingam. 2014. Prevalence of self-medication practices and its associated factors in Urban Puducherry, India. *Perspectives in Clinical Research* 5 (1): 32. https://doi.org/10.4103/2229-3485.124569.
Shirleyana, S., S. Hawken, and R. Y. Sunindijo. 2018. City of Kampung: Risk and resilience in the urban communities of Surabaya, Indonesia. *International Journal of Building Pathology and Adaptation* 36 (5): 543–568. https://doi.org/10.1108/IJBPA-02-2018-0025.
Subashini, S., B. Kumar Garla, M. Karuppaiah, and T. Taranath. 2017. Prevalence of self-medication practice among people attending oral health outreach programmes in Madurai East, Tamil Nadu. *Journal of Advanced Oral Research* 8 (2): 14–20. https://doi.org/10.1177/2229411217729104.
Susheela, F., N. Gorntla, P. Bhupalam, K. Veerabhadrappa, B. Sahithi, and S. M. Ishrar. 2018. Social support and self-care behavior study. *Journal of Education and Health Promotion* 7: 1–6. https://doi.org/10.4103/jehp.jehp.
Waljee, A. K., M. A. M. Rogers, P. Lin, A. G. Singal, J. D. Stein, R. M. Marks, J. Z. Ayanian, and B. K. Nallamothu. 2017. Short term use of oral corticosteroids and related harms among adults in the United States: Population-based cohort study. *BMJ (Clinical Research Edition)* 357: j1415. https://doi.org/10.1136/bmj.j1415.
WHO. 1998. *The role of the pharmacist in self-care and self-medication*. Department of Essential Drugs and Other Medicines, WHO. https://apps.who.int/iris/bitstream/handle/10665/65860/WHO_DAP_98.13.pdf;jsessionid=545DACF4F099BE0FB83AB2A855AE573E?sequence=1
Xu, J., X. Wang, K. S. Sun, L. Lin, and X. Zhou. 2020a. Parental self-medication with antibiotics for children promotes antibiotic over-prescribing in clinical settings in China. *Antimicrobial Resistance and Infection Control* 9 (1): 1–8. https://doi.org/10.1186/s13756-020-00811-9.
Xu, W., Z. Li, Z. Pan, R. He, and L. Zhang. 2020b. Prevalence and associated factors of self-treatment behavior among different elder subgroups in rural China: A cross-sectional study. *International Journal for Equity in Health* 19 (1): 1–12. https://doi.org/10.1186/s12939-020-1148-2.
Zollner-Schwetz, I., and R. Krause. 2015. Therapy of acute gastroenteritis: Role of antibiotics. *Clinical Microbiology and Infection* 21 (8): 744–749. https://doi.org/10.1016/j.cmi.2015.03.002.

Chapter 8
Assessing Rural Sustainability in Nangapanda, East Nusa Tenggara, Indonesia

Gabriel Andari Kristanto, Halvina Saiya, Dini Kemala, and Sanca Pamungkas

Abstract This chapter presents a comprehensive assessment of Nangapanda District, located in Flores, East Nusa Tenggara Province, Indonesia, with the objective of evaluating its sustainability. Drawing on data from 347 household questionnaires, the study examines various socio-economic indicators, housing conditions, access to electricity, and sanitation, covering access, types, infrastructure, usage, and quality. In addition, the study includes direct observations to assess the quality of clean water sources, waste management practices, and the cleanliness of sanitation facilities. To evaluate child health and nutritional status, the research analyzes data from 1,130 children, comprising 995 elementary school students across 24 schools and 175 preschool children aged 2–4 years. These findings are further integrated with interviews with parents and caregivers to ensure data reliability and contextual depth. The analysis identifies both strengths and persistent challenges across key sectors that influence the district's sustainability. Despite significant ongoing improvements, the district continues to encounter several obstacles that require attention. The chapter underscores the importance of coordinated government intervention and active multi-stakeholder engagement to enhance the long-term sustainability and well-being of Nangapanda District.

Keywords Clean water source · Nangapanda district · Rural sustainability · Sanitation · Socio-economic condition · Waste management

G. A. Kristanto · D. Kemala
Department of Civil Engineering, Faculty of Engineering, Universitas Indonesia, Depok, Indonesia
e-mail: andari@eng.ui.ac.id

H. Saiya
School of Environmental Science, Universitas Indonesia, Jakarta Pusat, Indonesia

S. Pamungkas (✉)
Department of Literary Studies, Faculty of Humanities, Universitas Indonesia, Depok, Indonesia

M. Budianta et al. (eds.), *A Liveable Kampung*, Engaging Indonesia,
https://doi.org/10.1007/978-981-96-7866-2_8

8.1 Introduction

In the late 1990s, many developing countries, particularly in the Asian region, experienced rapid urbanization and industrial expansion. This period of growth was accompanied by significant transformations and structural changes in rural areas (Ma et al. 2019). The shift from agrarian livelihoods to urban-industrial economies led rural populations to gradually abandon their traditional environments, often resulting in a decline in their overall quality of life (Li et al. 2019). By 2012, it was estimated that approximately 76% of poverty in developing countries was concentrated in rural areas, limiting access to sufficient and nutritious food (Ji et al. 2019; Giovanucci et al. 2012). In African nations, rural communities frequently struggle to meet basic food needs due to limited feedstock availability, low technological capacity, and poor agricultural productivity. Moreover, rural populations face significant challenges in accessing essential resources such as clean water, energy, and arable land (Khaleel and Ngah 2013). According to the World Health Organization (WHO), stark disparities remain in access to safely managed drinking water, available to 85% of urban residents compared to only 53% of rural populations. Sanitation access also lags behind, with only 43% of rural communities benefiting from adequate sanitation facilities (WHO 2017). Similarly, disparities in educational opportunity persist. Many rural residents are unable to complete formal education and, in some cases, do not attend school at all. These communities often grow within large households, further compounding socioeconomic vulnerabilities (UNDP 2017).

Despite the numerous challenges faced by rural areas, villages play a pivotal role in achieving a sustainable and resilient global society (Lekic et al. 2018; Sallu et al. 2010). Accordingly, individuals and rural communities are expected to demonstrate the capacity to adapt and transform in response to social, economic, environmental, and political challenges (Agboeze and Nwankwo 2018; Baldwin et al. 2017). This adaptive potential underscores the importance of sustainability as a guiding paradigm for development aimed at improving global living standards. With the adoption of the Sustainable Development Goals (SDGs) in 2015, rural populations have been increasingly recognized as key factors in inclusive development. The SDG framework promotes equity and justice for rural communities by expanding access to essential services such as education, healthcare, economic opportunities, legal protection, public safety and security (United Nations 2015).

As the fourth most populous country in the world, with a population of approximately 284 million, Indonesia had an estimated 43.3% of its population residing in rural areas (Badan Pusat Statistik Indonesia/BPS 2025). While the national poverty rate has shown a steady decline, from 10.12% in 2017 to 9.66% in 2018 (Dib et al. 2018), certain provinces, particularly in eastern Indonesia, continue to experience disproportionately high levels of poverty. In rural areas, the incidence of poverty remains more pronounced, with East Nusa Tenggara (NTT) standing out as a critical case. In 2018, the rural poverty rate in NTT reached 24.65%, placing it as the fourth highest among all provinces in Indonesia (Badan Pusat Statistik Indonesia/BPS 2018). Additionally, the Village Development Index in NTT remains relatively low, recorded at under 59%, with basic service coverage at 49.54% and infrastructure quality at 28.96% (Badan Pusat Statistik Indonesia/BPS 2018). These indicators highlight ongoing development challenges and underscore the importance of targeted interventions to improve rural livelihoods and infrastructure in the region.

Various studies on sustainability have been conducted across both developed and developing countries. Several qualitative and quantitative factors have been identified as contributing to the sustainability of a region. These include anthropogenic elements (such as technology and development), social dimensions (including social capital, social infrastructure, and standards of living), economic variables (such as economic diversification, business activity, and human resource quality), and ecological considerations (such as environmental conditions and impacts) (Gorbenkova et al. 2018). Rodriguez-Pose and Hardy (2015) emphasized that in developing countries, quality of life and access to rural natural resources are closely tied to the availability of essential infrastructure, including water, wastewater management, and energy systems, as well as public facilities such as education and healthcare. Additionally, mobile infrastructure for the movement of people and goods further supports rural sustainability (Jha et al. 2017; Adisa 2012). Other studies have highlighted the importance of sustainable housing and household development strategies in rural areas, emphasizing the central role of housing in reinforcing cultural identity and social cohesion within rural communities (Amado et al. 2017; Gao et al. 2017).

This study aims to evaluate the sustainability of Nangapanda Subdistrict (Kecamatan Nangapanda), located on Flores Island in East Nusa Tenggara Province, Indonesia, using a set of indicators commonly applied in global sustainability assessments. The relevance of this research lies in the geographic and environmental context of the study area. Indonesia, situated within the Pacific Ring of Fire, is highly vulnerable to various natural disasters, including geophysical hazards such as earthquakes, tsunamis, landslides, and volcanic eruptions, as well as hydrometeorological events such as floods and droughts (Parwanto and Oyama 2014). One of Indonesia's most significant earthquakes occurred in East Nusa Tenggara in 1992, registering a magnitude of 7.8 and triggering a devastating tsunami (Sinki et al. 2021). The region has since experienced recurring seismic activity, resulting in ongoing damage and loss of life. In addition to seismic hazards, East Nusa Tenggara frequently faces crises linked to drought and environmental degradation, further underscoring the urgency of sustainable development in the area. Drawing on existing literature, this study developed a set of sustainability indicators, organized into eight thematic categories, to comprehensively assess the sustainability of rural areas (Boliko and Ialnazov 2019). The research employed a mixed-methods approach, incorporating literature review, structured questionnaires, interviews, and field observations to collect and cross-validate data.

8.2 Materials and Methods

8.2.1 Study Area

Nangapanda Subdistrict is situated in East Nusa Tenggara Province, Indonesia, on Flores Island—a small yet strategically significant island located approximately 1700 kilometers straight-line east of the capital city, Jakarta. The area is accessible via a four-hour flight from Jakarta to the nearest airport in Ende City. From Ende, travelers typically continue by land along the scenic coastal road for approximately 25 kilometers, a journey that takes around one and a half hours by car (Fig. 8.1).

In 1992, the island of Flores experienced a massive earthquake with a magnitude of 7.8 on the Richter scale centered at a depth of 35 km northwest of Maumere, the capital of Flores. The earthquake triggered a tsunami with a wave height reaching up to 35 m; resulting in over 1000 fatalities and widespread structural and buildings damage (Parwanto and Oyama 2014). The tsunami was not the first one to impact the island; historical records indicate a significant tsunami in Flores also occurred in 1928. These events reflect the island's proximity to the tectonic convergence zone between the Indo-Australian and Eurasian plates (Sarmiento et al. 2019).

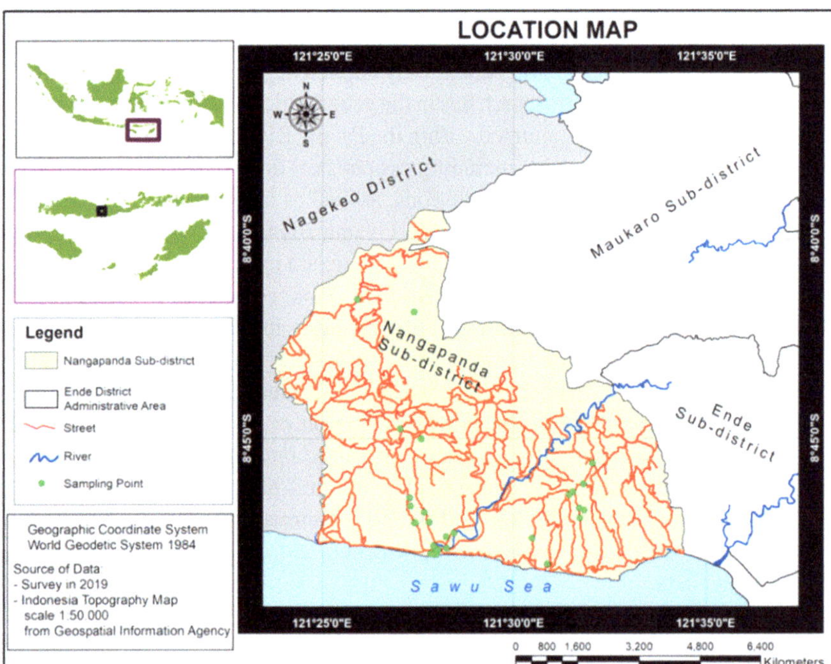

Fig. 8.1 Location map of the study area in Nangapanda Subdistrict, Flores Island, East Nusa Tenggara Province, Indonesia.

Nangapanda District consists of 29 subdistricts administrative units, including 28 villages (desa) and one urban ward (kelurahan), covering a total area of 213.17 km^2. As of 2017, the district had a population of 21,179 residents. It experiences a relatively dry tropical climate, with average temperatures ranging from 22 °C to 33 °C. Nangapanda accounts for 10.42% of the total area of Ende Regency, making it the largest subdistrict within the regency. Geographically, it is a coastal region with highly varied topography: 0.71% of the area lies at 0–25 meters above sea level (masl), 15.5% at 25–100 masl, 50.91% at 100–500 masl, and 32.83% above 500 masl. This mountainous condition poses logistical and transportation challenges particularly for rural and inland villagers.

According to the East Nusa Tenggara Regional Development Work Plan (RKPD), Nangapanda Subdistrict is designated as part of Development Area 4, with green stone (zeolite) identified as a key regional resource (Pemerintah Kabupaten Ende, 2015). These zeolite deposits, found along the coastal areas of Nangapanda, serve as a source of income for local residents through collection and sale.

8.2.2 Data Collection, Survey Instrument Design and Testing

Prior to the implementation of this study, ethical approval was granted by the Ethics Committee of the Faculty of Medicine, Universitas Indonesia. Data collection involved a combination of structured questionnaires, interviews, and direct physical examinations. The questionnaires captured comprehensive information on village-level socio-economic conditions, including demographics, education, employment, expenditures, and household characteristics. Housing conditions, such as ownership status and structural attributes, were also recorded. Additional sections addressed health indicators, including children's academic ability and nutritional status, sanitation (access, infrastructure types and use, and quality), sources of clean water, waste management, and electricity access. Data obtained through direct inspection focused on the quality of sanitation facilities, including clean water sources, waste composition, and the cleanliness of sanitation infrastructure. The variables, indicators, and sustainability outcomes are summarized in Table 8.1. The primary respondents for household-level data, covering housing, electricity access, sanitation, and waste disposal, were heads of households and their spouses, drawn from a total of 347 households. For the health component, the study assessed the nutritional status of children attending elementary schools and preschools in Nangapanda District. Anthropometric indicators such as age, weight, and height were used to determine nutritional conditions, including underweight, overweight, wasting, and anemia. In total, 1,130 children participated in the assessment, comprising 995 students from 24 elementary schools and 175 preschool children aged 2–4 years.

To evaluate academic ability, the study included 188 child–parent/caregiver pairs. After administering the questionnaires, researchers conducted interviews with parents and caregivers to understand their perceptions of the importance of education. The study also assessed water quality by collecting 21 samples from springs, reservoirs, and taps located across eight sites within Nangapanda Subdistrict. Analyses included physical, chemical, and biological parameters. Water testing followed the Hach Water Analysis method, approved by the U.S. Environmental Protection Agency for physical and chemical parameters, and the APHA Method 9221 for total and fecal coliform determination. Waste management practices were described qualitatively, while quantitative measurements of waste generation were carried out specifically along the coastal area of Ende.

Secondary data were obtained from the Central Statistics Agency (Badan Pusat Statistik, BPS), particularly the publication Nangapanda in Figures. Socio-economic variables reviewed included demographic composition, educational attainment, and employment patterns. Supplementary data from BPS (2019) provided detailed information on population by gender and the distribution of educational facilities in Nangapanda District. Electricity access was analyzed using indicators related to electricity sources, including PLN (state-owned electricity), non-PLN sources such as solar power (PLTS), and other alternative systems.

Prior to fieldwork, all enumerators received comprehensive training, which included data collection procedures, survey instrument familiarization, geographic orientation, pre-testing of survey tools, household sampling techniques, and reporting protocols. Field supervisors conducted regular monitoring and oversight throughout the data collection process to ensure methodological rigor and high data quality.

Table 8.1 Key variables, indicators, and positive sustainability outcomes of the study

Variables	Indicators	Sustainability results
Household characteristics	Age, sex, source of income, and monthly income	Percentage of household head having better socio-economic indicators
Housing characteristics	Different housing structures, sharing of house/room	Percentage of household having affordable housing facility
Access to education	Facilities, infrastructure, distance, poverty/economic condition	Percentage of household head having higher education level
Access to electricity	Electricity facility	Proportion of household connected with electricity
Access to sanitation and waste disposal	Toilet, septic tank	Percentage of household using basic adequate sanitation; waste management system
Children educational status	School enrollment, ability to pay	Percentage of children who do not drop out of school
Water quality	Physical, chemical, and E. coli and total coliforms bacteria	Percentage of water source with good water quality

8.3 Sustainability of Villages in Nangapanda, Indonesia

8.3.1 Village Characteristics

8.3.1.1 Demography

Figure 8.2 illustrates the total population of men and women in Nangapanda District in 2018. During that year, the district recorded a total population of 22,859, comprising 11,339 males and 11,520 females. Population density varied significantly across the 29 villages, ranging from fewer than 21 to 251 inhabitants per square kilometer. Population growth trends also differed by village. While several villages experienced positive growth, others reported negative growth rates. Tendambepa Village recorded the highest annual growth rate at 9.6%, whereas Tanazozo Village experienced the lowest, with a decline of −10.4%. On average, the annual population growth rate in Nangapanda District in 2018 stood at 2.7%.

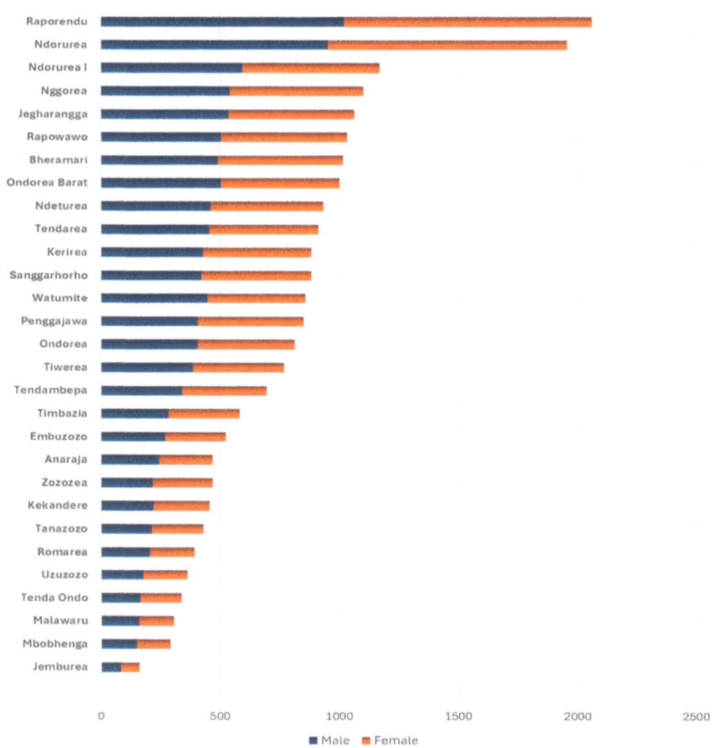

Fig. 8.2 Population distribution by gender across villages in Nangapanda Subdistrict

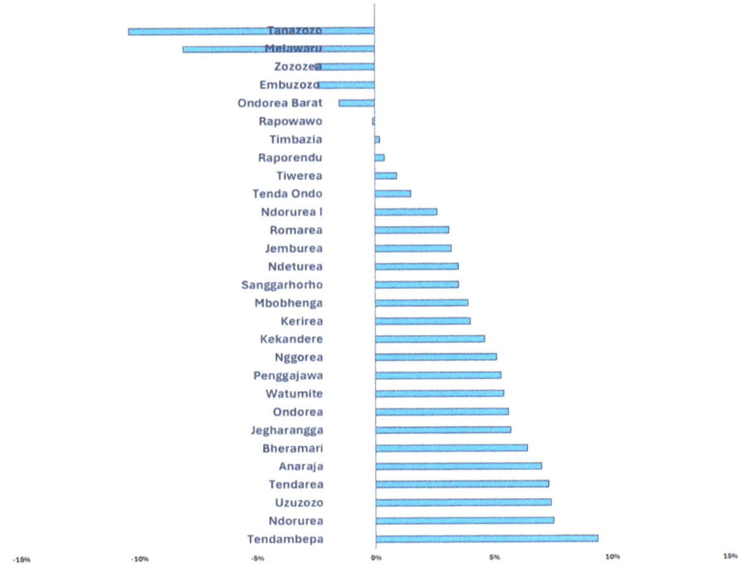

Fig. 8.3 Population growth rates by village in Nangapanda Subdistrict

Several factors including migration and fertility rates, influence population growth in rural areas (Liu et al. 2017). However, when population growth is not accompanied by an increase in public facilities and employment opportunities, it can lead to a decline in the community welfare and economic conditions (Straka and Tuzova 2016). This issue is particularly significant given that individuals of productive age group represents the largest composition in the population. Therefore, changes in village population need close attention in relation to the potential economic and social impacts (Figs. 8.3, 8.4).

8.3.1.2 Education

Indonesia has undergone a significant trajectory of educational development, beginning in the New Order era (1967–1998) with the implementation of nine-year compulsory education, the expansion of private educational institutions, and the adoption of a decentralized system at the primary and secondary levels (Kristiansen and Pratikno 2006). Despite these advances, disparities in educational access between urban and rural areas persist, particularly at the secondary education level. Previous studies have reported that junior high school participation rates in Indonesia remained about 60%, a condition predominantly affecting regions outside of Java, especially in eastern Indonesia (Kristiansen and Pratikno 2006).

Fig. 8.4 Schoolchildren participating in a community clean-up activity

A similar pattern is observed in Nangapanda Subdistrict, where a total of 4,104 children were enrolled in school. The student distribution consisted of 61% in elementary school, 28% in junior high school, and 11% in senior high school. Educational facilities in the subdistrict are heavily concentrated at the elementary level, with 38 elementary schools, compared to only 8 junior high schools and 2 senior high schools. Academic test results for Grade 1 and 2 elementary students indicate that 21.8% of children performed above average, 62.2% were at average level, and 16% scored below average. However, most schools in the 21 surveyed villages lacked adequate learning infrastructure.

Students typically received only one or two school-provided textbooks, distributed free of charge. Moreover, the availability of learning facilities remains insufficient relative to student needs; such as inadequate desks limiting the effectiveness of the teaching and learning process (Mughal et al. 2019).

8.3.1.3 Occupation

Rural communities continue to rely significantly on the agricultural sector to sustain their livelihoods and support the local economy (Rustinsyah 2019). According to the World Development Indicator, agriculture serves as a key driver in reducing rural poverty, providing income for nearly 40% of the population. The Food and Agriculture Organization (FAO) underscores that agricultural growth is twice as effective as growth in other sectors in alleviating poverty and hunger, particularly in low-income regions (FAO 2018). Consequently, agriculture plays a critical role in advancing poverty alleviation initiatives.

In Nangapanda, the study found that 63.5% of household work as farmers. Additionally, 35.6% are employed in non-agricultural sectors, while only 0.9% reported being unemployed. The study also revealed that women actively contribute to the household economy. Based on questionnaire responses, 50.6% of wives engage in income-generating activities, with many working in agriculture, while 49.4% do not participate in the labor force (Gignoux and Menendez 2016).

The non-agricultural sector in Nangapanda features a wide range of occupations, including fishermen, civil servants, private employees, laborers, entrepreneurs, traders, and Indonesian migrant workers. This occupational diversity contributes to varying household income levels, ranging from Rp 2,000,000 to Rp 7,000,000 per month (2025 exchange rate: 1 USD = Rp 16,200). Expenditure patterns show that the largest portion of household budgets, an average of 43.05%, is allocated to primary needs such as food, followed by 7.92% for other essential household items. This spending structure leaves limited room for discretionary expenses and savings.

8.3.2 Housing Characteristics

The concept of sustainable housing encompasses multiple dimensions. Perrucci (2016) identifies key aspects such as affordability , energy efficiency, the use of appropriate building materials, and structural safety. In developing countries, especially in earthquake-prone regions, the choice of building materials are very important, particularly due to increased vulnerability to natural disasters (Gan et al. 2017). The major earthquake and tsunami in Flores along with subsequent smaller seismic events, have underscored the urgent need for resilient housing structures. These experiences highlight the importance of designing buildings that prioritize structural integrity to minimize casualties and protect communities.

Building material play a critical role in housing construction, particularly in earthquake-prone regions. Houses constructed with lightweight materials generally face a lower risk of structural damage during seismic events compared to those made with heavier materials (Ahmed and Morita 2018). Heavier materials such as bricks tend to be rigid and lack elasticity, making them suitable only for withstanding minimal ground movement. Moreover, due to their substantial weight, brick structures require significant inertial resistance and additional wall reinforcement to withstand the tensile stresses generated by earthquakes (Baghel 2019). Observations of housing structures in Nangapanda reveal that most homes are composed of three key elements: roof, walls, and flooring (Table 8.2). Natural materials such as wood and bamboo are common in house construction (Fig. 8.5). Approximately 95% of roofs are supported by wooden structures, while concrete and mild steel roofs are used much less frequently—only 1.5% and 1.2%, respectively. In terms of wall construction, the dominant materials include brick (39.7%), bamboo (19.7%), and wood (17.1%). For flooring, 64.4% of households use plastered floors, indicating a shift toward more permanent materials. In particular, the increased use of brick as a primary wall material is especially evident in houses constructed within the past decade.

The use of bricks in house construction without proper structural reinforcement design often proves inadequate during earthquakes due to the material's inherent weakness, lack of elasticity, and brittleness. In contrast, bamboo is a lightweight material with high tensile strength, making it highly flexible and capable of withstanding high wind speeds and earthquakes (Baghel 2019). An additional advantage of bamboo in disaster-prone regions is the relatively short repair time required for bamboo structures compared to those made of brick (Baghel 2019). However, several considerations must be addressed when selecting bamboo as a material. The durability is generally shorter (30–40 years) compared to other materials such as bricks and concrete. Therefore selecting the appropriate type of bamboo species and ensuring proper maintenance are critical. In addition, extreme heat and climatic conditions may affect the thermal performance of bamboo walls. In tropical climates, applying a mud coating to bamboo walls can help reduce heat transfer from the external environment to the interior, thereby improving indoor thermal comfort (Baghel 2019).

The limited availability of building materials poses significant challenge especially for rural communities or low-income households in constructing affordable housing, particularly due to cost-related constraints (Omer and Noguchi 2020). The use of locally available materials such as wood, sand, clay, lime, etc., has significantly contributed to reducing construction costs, with potential saving of up to 60% (Ugochukwu and Chioma 2015). This approach enables rural communities to access adequate housing while fostering resilient local development (Omer and Noguchi 2020; Celentano et al. 2019). Previous research has shown that the majority of tsunamis in Indonesia are triggered by earthquakes (Parwanto and Oyama 2014). To minimize casualties, it is essential to ensure the availability of early warning systems and to enhance community awareness and preparedness. Equally important is the implementation of standardized building practices and government support in constructing earthquake-resistant housing that adheres to established earthquake safety standards.

In addition to building materials, household size plays a crucial role in ensuring comfort and privacy within the home. The concept of overcrowded housing is understood both quantitatively, as the density of occupants per unit area, and qualitatively, as individuals' psychological responses to limited space (Royuela et al. 2019). The study found that the majority of houses in Nangapanda are inhabited by 4–6 individuals (64.8%), while 9.8% accommodate only 1–3 occupants. In contrast, 22.2% of homes are shared by 7–10 people, and 3.2% are occupied by more than 10 individuals. Royuela et al. (2019) observed that overcrowding is commonly found in rural areas, a trend that is also evident in Nangapanda, where some homes are inhabited by multiple families. Contributing factors include cultural norms, high birth rates, and the presence of adult family members who remain financially dependent (Pablo and Javier 2016). Addressing these factors is essential for improving housing quality and promoting well-being in rural communities reported cases of overcrowding often found in rural areas. This condition is also found in Nangapanda district where there

Table 8.2 Housing characteristics in Nangapanda Subdistrict

	Percentage (%)
House ownership	
Privately owned house	87.9
Contracted house	0.3
Rented house	0.9
Rent-free house	1.7
Relative-owned	9.2
Rent-free house	
Housing structure	
Roof	
Concrete	1.5
Wood	95
Bamboo	2.1
Light Steel	1.2
Others	0.4
Wall	
Hebel	1.5
Plastered	4.4
Red brick	39.7
Painted	14.7
Wire	0.3
Wood	17.1
Plywood	5.3
Bamboo	19.7
Floor	
Marble	0.6
Ceramic	21.5
Tile	1.8
Plastered	64.4
Wood	1.5
Parquetted	1.2
Bamboo	3.2
Soil	5
Others	1.3
Number of household member	
1–3 person	9.80
4–6 person	64.8
7–10 person	22.2
> 10 person	3.2

are several houses inhabited by more than one family. Some factors such as culture, high birth rate and the presence of adult family members who are not yet financially independent are the factors causing the number of house occupants not proportional to their capacity (Pablo and Javier 2016).

8.3.3 Children's Nutritional Status

Nutritional status, reflected in both the quantity and quality of food intake, serves as a key determinant of quality-of-life standards (Torlesse et al. 2016). According to UNICEF, an estimated one in three children under the age of five worldwide experience growth disruption due to malnutrition, most commonly in the form of stunting, wasting, or overweight (UNICEF 2019). The prevalence of overweight and obesity among children aged 5–19 years increased significantly to one in five children (UNICEF 2019).

The nutritional status assessment of school-aged and preschool children in Nangapanda focused on indicators such as underweight, overweight, wasting, and stunting. Among elementary school children, 27.3% were classified as underweight, with 9.2% falling into the severely underweight category. Thinness was observed in 16.1% of children, while 3.6% were classified as severely thin. Additionally, the prevalence of stunting and severe stunting was 22.4% and 3.2%, respectively. In preschool-aged children, 20% were underweight and 5.2% were severely underweight. Wasting and severe wasting were recorded at 6.2% and 1.4%, respectively. Stunting affected 22.4% of preschoolers, while 7.1% experienced severe stunting. Blood test results further revealed a high prevalence of anemia, with 30.4% of school-aged children and 33.2% of preschool-aged children affected. These findings highlight significant nutritional and health challenges among children in the district, indicating the need for targeted nutrition and health interventions.

Stunting in early childhood is closely associated with cognitive, mental, and physical developmental impairments, which can negatively impact educational attainment, productivity, and future income potential (Baye 2017; Canavan et al. 2016). Baye (2017) notes that individuals who experience inadequate nutrition during their growth years are likely to exhibit reduced work performance, which can ultimately impact national economic productivity. Historical trends and projections of stunting suggest that its persistence could reduce Gross Domestic Product (GDP) by as much as 6% by 2050 (Horton and Steckel 2013). In addition, the consequences of poor nutrition are often irreversible, particularly when they occur during critical periods of growth and development. Several factors influence a child's nutritional status. A study on childhood stunting in Indonesia found that household economic conditions, specifically household poverty, accounted for approximately 93% of stunting prevalence. Poverty significantly impacts a child's access to adequate nutrition and healthcare, increasing the risk of stunting. In addition, access to clean water, the quality of water sources, and household environmental sanitation contributed 4%, 9%, and

4%, respectively (Rizal and van Doorslaer 2019). These findings highlight the multifactorial nature of stunting and emphasize the importance of integrated interventions that address both economic and environmental determinants of nutrition.

8.3.4 *Access to Sanitation and Clean Water*

8.3.4.1 Clean Water Access

Amid the growing demand for clean water and sanitation, progress in expanding access remains uneven across countries, particularly among rural communities and vulnerable populations (Ezbakhe et al. 2019). According to the World Health Organization (WHO), access to safe drinking water is measured by the proportion of the population using improved water sources, such as municipal piped water connections (PAM/IPAM), protected dug wells, protected springs, and rainwater harvesting systems.

Figure 8.5 titled Distribution of Clean Water Sources Used by Households in Nangapanda Subdistrict, shows 40.3% of households relying on springs, 33.2% on water supplied by the municipal water utility company (PAM/PDAM), and 21.8% on well water. Overall, springwater, municipal waters and wells serve as the primary sources of clean water in the district. Despite the relative abundance of springs, access remains limited for some households due to factors such as remote locations and challenging topography. Furthermore, not all spring water is distributed via the PAM piping network, requiring residents to manually collect water using plastic containers. The volume transported is limited and constrained by individual physical strength, which complicates the ability to meet daily water needs, particularly during the dry season as shown by schoolchildren carrying plastic containers (Fig. 8.4). In several highland villages, where access to spring water is especially limited, residents rely on rainwater collected in personal storage tanks with very limited capacity or through rainwater harvesting systems (Fig. 8.6).

In 2017, World Health Organization (WHO) reported that 82% of rural communities in Indonesia had access to improved drinking water (WHO 2017). The study results indicate that access to clean water in Nangapanda stands at 77.4%, which is below the national village average of 82% as reported by the WHO. This gap highlights a critical area for improvement in ensuring equitable water access.

The availability of clean water sources is closely linked to a region's resilience and vulnerability in the context of climate change (Howard et al. 2010; Howard and Bartram 2010). Limited access increases the risk of water insecurity, particularly during prolonged dry seasons, and weakens the community's capacity to adapt to environmental stressors. In this context, the quality of infrastructure and management systems for clean water sources plays a critical role in determining a region's capacity to withstand the impacts of climate change. In Ende Regency, where rainfall levels and intensity are relatively low and vary significantly across months and regions, the reliability of water access becomes a pressing concern. According to a prior

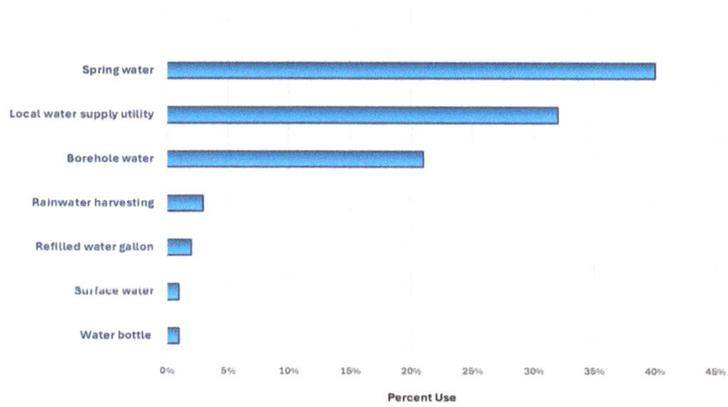

Fig. 8.5 Distribution of clean water sources used by households in Nangapanda Subdistrict

Fig. 8.6 Innovative housing design in Nangapanda featuring integrated rainwater reservoirs

assessment of climate resilience, the water supply system in the area is categorized as low to medium (Howard and Bartram 2010). Ende Regency experiences a relatively dry tropical climate, with average temperatures ranging from 22 °C to 33 °C, further emphasizing the importance of effective water resource management in enhancing regional resilience.

In addition to issues of quantity and continuity, the quality of clean water in Nangapanda District is compromised by bacterial contamination in several communal storage facilities. This study found that 30% of water reservoirs contained fecal coliform and total coliform bacteria, exceeding the threshold established by the Indonesian Minister of Health Regulation No. 492 of 2010, which mandates a zero count per 100 mL for both types of bacteria in clean water. Such contamination has been identified as a contributing factor to recurrent diarrhea outbreaks, particularly during seasonal transitions between the dry and rainy seasons (Patunru 2015). The highest incidence of diarrhea was recorded in Ndorurea and Raporendu villages, which also showed elevated concentrations of Escherichia coli in water sources. In these areas, local farming practices, specifically, allowing livestock to roam freely, have led to fecal contamination of the environment. Rainwater runoff frequently transports animal waste into surrounding water sources, including springs used for household consumption. Although local public health centers (puskesmas) are aware of these conditions, the disinfection of communal water reservoirs is not performed regularly, further contributing to the outbreaks of waterborne diseases.

Restoration and conservation represent viable solutions for improving access to clean water sources in Nangapanda. Given the unpredictable rainy season and the low rainfall frequency and challenging topography, therefore implementation of appropriate technologies and simple infrastructure is essential to ensure the continuity of both existing and potential water sources (Sen and Kansal 2019). Examples of such interventions include the construction of swales, artificial ponds, and filter ponds, strategies that have been successfully applied in rural settings worldwide (Rujner et al. 2018; Øhlenschlæger et al. 2016). Swales, in particular, are effective in enhancing water infiltration and reducing runoff and pollutants, while requiring relatively low installation and maintenance costs (Rujner et al. 2018). In addition, the protection of natural water sources remains critical to prevent contamination and maintain water quality for community use.

Despite these promising approaches, improving the water supply system in Nangapanda faces several challenges. Interviews with key community stakeholders revealed that public participation in addressing clean water issues remains limited, with only 47.5% of residents expressing willingness to engage in water supply initiatives. Nevertheless, government programs such as PAMSIMAS (Community-Based Drinking Water and Sanitation Provision) have made significant contributions to infrastructure development in the region (Fig. 8.6). One notable example is the development of the Aekome spring in Tanarhi Village, where three new reservoirs were constructed. The design of the pipeline system was based on a 15-year projection to serve the villages of Tanarhi, Mboturea, and Bhoawora, aiming to meet the clean water needs of approximately 667 residents. Additionally, the reservoirs are equipped with simple filtration systems designed to reduce dissolved solids in the water, enhancing its suitability for household use (Fig. 8.7).

Fig. 8.7 Integration of natural spring water with an engineered water distribution system

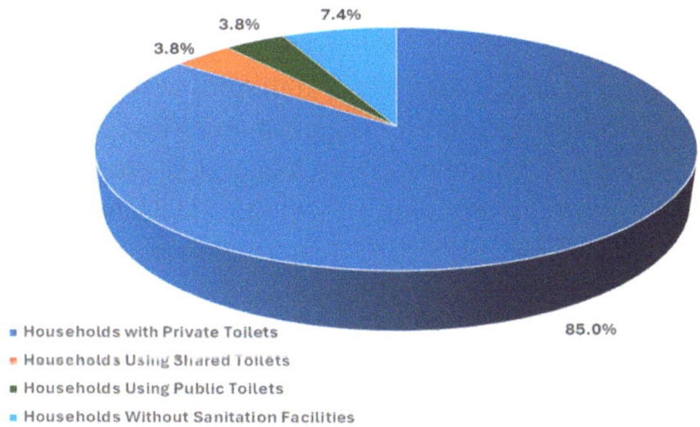

Fig. 8.8 Household toilet access in Nangapanda Subdistrict

8.3.5 *Sanitation*

According to the World Health Organization (WHO), access to adequate sanitation is measured by the percentage of the population using improved sanitation facilities—those that hygienically separate human excreta from human contact. These include toilets connected to wastewater treatment systems, septic tanks, pour-flush latrines, ventilated improved pit latrines, and pit latrines with slabs. Other types, such as shared or unimproved sanitation facilities, are not considered adequate under this definition.

In recent years, Ende Regency has shown notable progress in expanding access to sanitation facilities. In 2017, 64.98% of households had access to private toilets, increasing to 79.25% by 2018 (BPS 2019). The findings of this study further indicate that 84.1% of households now have access to private toilet facilities (Fig. 8.8), while 3.8% use shared toilets, another 3.8% rely on public toilets, and 7.4% of households lack access to any toilet facility. These figures suggest that sanitation coverage in Nangapanda exceeds the national rural average. However, observations reveal that the mere presence of toilets does not ensure their regular or hygienic use. Limited availability of clean water remains a key barrier, often preventing consistent toilet use and maintenance, thereby undermining overall sanitation outcomes.

8.3.6 *Waste Management*

Effective waste management plays a critical role in advancing the Sustainable Development Goals (SDGs), particularly in promoting sustainable cities and communities, responsible consumption and production, clean water, and adequate sanitation. In addition, sound waste management practices contribute to climate change mitigation by preventing and reducing greenhouse gas emissions (Rodic and Wilson 2017).

From an economic standpoint, the waste sector holds significant potential to generate employment and support inclusive and sustainable economic growth (Wang et al. 2018a, b).

Despite these opportunities, the waste sector continues to face considerable challenges, particularly in developing countries (Apostle and Mhai 2012; Sudibyo et al. 2017). Globally, an estimated 2 billion people lack access to proper waste collection services, resulting in widespread dumping in uncontrolled landfills or open environments (Li et al. 2018; UNEP 2015). In Mexico, approximately 96% of waste is managed through open burning (Reyna-Bensusan et al. 2018) while in China, between 30% and 60% of waste is indiscriminately discarded (Wang et al. 2018a, b). Contributing factors include low public awareness, inadequate facilities, and insufficient regulatory enforcement, particularly in rural and peri-urban settings (Apostle and Mhai 2012; Tadesse et al. 2008; Drahansky et al. 2017).

Similar challenges persist in Nangapanda Subdistrict where the lack of formal solid waste management infrastructure compels residents to handle waste independently, often through unsustainable means. A significant majority of households (93.2%) dispose of waste through open burning, while 8.1% discard waste into rivers, creeks and the sea, practices that lead to coastal pollution and contribute to the gradual degradation of marine reef ecosystems. Additionally 4% of households engage in indiscriminate dumping, further exacerbating environmental challenges in the region (Fig. 8.9). In contrast, only 0.6% of households compost their waste, and just 1.2% utilize official waste collection services. These findings highlight the pressing need for community-based waste management solutions and infrastructure development to support environmental sustainability in the region.

Open burning of household waste poses significant risks to both environmental quality and public health. Cogut (2016) reported that the global burning of approximately 1.4 trillion kilograms of waste annually contributes around 5% of total CO_2 emissions, while the burning of 3.6 billion kilograms per year accounts for 1% of global CH_4 emissions. Exposure to pollutants generated from open burning is

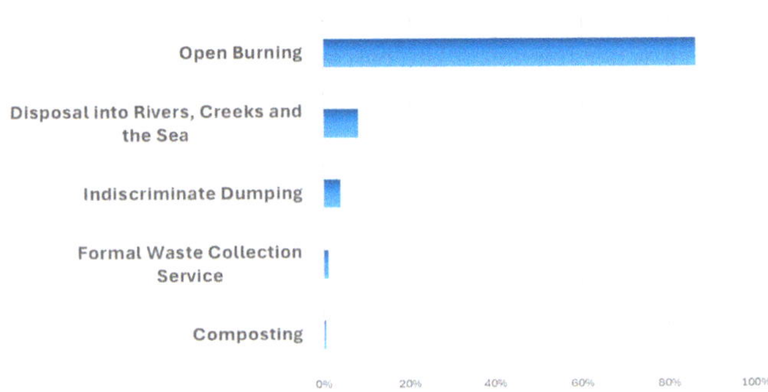

Fig. 8.9 Distribution of household waste disposal methods in Nangapanda Subdistrict

associated with various health issues, including respiratory disorders, skin irritation, and increased risk of lung cancer (Cogut 2016). In Nangapanda Subdistrict, waste-related challenges extend beyond air pollution, with significant accumulation of inorganic waste, particularly plastic debris, along the coastal areas. The subdistrict's coastal location exposes it to a high influx of marine debris carried by ocean currents, leading to widespread beach pollution. A cleanup of a 100-meter stretch of shoreline collected approximately 18.75 kilograms of accumulated waste, predominantly composed of plastic food and beverage packaging. Prolonged exposure to high temperatures accelerates the breakdown of plastic waste into microplastics, which pose enduring environmental and health threats. These microscopic particles can infiltrate soil, contaminate water sources, and accumulate in marine ecosystems. Once in the ocean, microplastics disrupt the food chain and elevate the risk of human exposure through seafood consumption. As seafood forms a substantial part of the local diet, the potential health implications are particularly concerning. Recent studies have linked microplastic ingestion and inhalation to gastrointestinal, respiratory, and endocrine disorders. These findings underscore the urgent need for enhanced coastal waste management and proactive pollution prevention strategies (Prata et al. 2020; Walkinshaw et al. 2020).

Addressing these waste management issues requires prioritization at the household level, particularly in both rural and urban settings (Moharana 2012). Community-level interventions become essential when waste cannot be adequately managed by individual households. Waste sorting remains a fundamental and accessible step in the waste management hierarchy. Moreover, promoting recycling can add value to waste materials and reduce environmental burdens. Home composting emerges as a viable strategy in rural communities, where the organic fraction of household waste is typically high (Sudibyo et al. 2017; Mihai et al. 2017). For inorganic waste with market value, the informal sector plays a key role in its collection and redistribution, offering both environmental and economic benefits. Ultimately, the success of sustainable waste management initiatives depends heavily on the active participation and engagement of local communities.

8.3.7 Electrical Access

Households in Nangapanda Subdistrict rely on various sources of electricity, including electricity supplied by the State Electricity Company (PLN), non-PLN sources such as solar power (PLTS), as well as traditional lighting methods like petromax lanterns and torches. As shown in Table 8.3 the vast majority of households (98.3%) utilize electricity from PLN, indicating strong grid coverage and access to formal electricity infrastructure. A smaller proportion of households rely on alternative sources: 0.3% use non-PLN electricity in the form of solar power systems (PLTS), another 0.3% use petromax lanterns, and 0.6% depend on torches for lighting. These findings highlight a high level of electricity access in the subdistrict, while

Table 8.3 Primary sources of household electricity in Nangapanda Subdistrict

Electrical sources	Percentage (%)
PLN-generated electricity	98.3
Non-PLN electricity (solar-powered)	0.3
Petromax lantern	0.3
Torch	0.6

also pointing to a small segment of the population that still depends on off-grid or traditional lighting solutions.

Access to electricity is a critical component in achieving the Sustainable Development Goals (SDGs), particularly in addressing energy poverty (Litzow et al. 2017). In terms of household coverage, electricity access in Nangapanda Subdistrict can be considered relatively high. However, evaluating the sustainability of energy supply, both electricity and fuel, requires a broader assessment that includes efficiency and the diversity of energy sources (UNDESA 2019). Field observations indicate that although nearly all households rely on electricity supplied by the State Electricity Company (PLN), which continues to dominate the national electricity sector, the reliability of supply remains a concern. Intermittent power outages still occur in certain areas and households, indicating instability in electricity distribution. In addition, the contribution of renewable energy sources in Nangapanda remains minimal, limiting the overall sustainability of the local energy system. Efforts to expand renewable energy-based electrification in rural areas must carefully consider the social and economic conditions of local communities, particularly in regions like Nangapanda that benefit from abundant solar energy potential. Prior research on rural electrification highlights that the success of such initiatives is strongly influenced by investment capacity, as well as the willingness and ability of residents in remote areas to pay for electricity. These factors play a critical role in determining the feasibility and sustainability of electricity access in rural Indonesia (Blum et al. 2013). These findings underscore the importance of inclusive, community-sensitive approaches to ensure not only widespread access but also sustainable and resilient energy systems.

8.4 Conclusion and Recommendations

This study investigates the multifaceted challenges rural areas face in progressing toward the Sustainable Development Goals (SDGs). The findings highlight several interrelated issues—spanning social, health, and environmental domains—that may hinder Nangapanda District's transition into a sustainable village. Among the key concerns are uneven population growth and limited access to educational facilities. Although public awareness of the importance of education has increased, economic hardship remains a primary driver of school dropouts. Additionally, a number of school-aged and preschool-aged children are affected by growth-related health problems such as stunting, wasting, and thinness, which are closely linked to inadequate nutritional status. Access to clean water, sanitation, and effective waste management also pose significant challenges. The presence of bacterial contamination, particularly Escherichia coli and total coliforms, has been detected in several water sources—most notably in communal reservoirs. These findings suggest a correlation with seasonal diarrhea outbreaks, especially during the transition from the dry to the rainy season. The lack of solid waste management infrastructure contributes to limited public knowledge about proper disposal practices and leads to a high incidence of open burning. Addressing these challenges requires coordinated intervention from multiple stakeholders, particularly government institutions. Although the majority of Nangapanda residents are employed, the wide range of income levels reflects disparities in the overall quality of life. The provision and enhancement of public facilities are essential to ensure equitable access to education and health services. In the case of clean water provision, community collaboration has proven instrumental in mitigating water-related issues. Through sustained efforts in infrastructure development, community engagement, and policy support, Nangapanda—and rural areas across Indonesia can strengthen their path toward sustainability, particularly in the face of emerging challenges.

Acknowledgements This study was supported by Direktorat Riset dan Pengabdian Kepada Masyarakat Universitas Indonesia (DRPM UI) under PINTERMIDI UI Research Grant No. 945/SK/R/UI/2019.

References

Adisa, R. S. 2012. Rural development in the twenty-first century as a global necessity. In *Rural development—Contemporary issues and practice*, ed. R. S. Adisa, 2–14. London: IntechOpen. https://doi.org/10.5772/49098.

Agboeze, M. U., and E. A. Nwankwo. 2018. Actualizing sustainable development goal-11 in rural Nigeria: The role of adult literacy education and tourism development. *Business Strategy and Development* 1 (3): 180–188. https://doi.org/10.1002/bsd2.21.

Ahmed, M. S., and H. Morita. 2018. An analysis of housing structures' earthquake vulnerability in two parts of Dhaka city. *Sustainability* 10 (4): 1106. https://doi.org/10.3390/su10041106.

Amado, M. P., I. Ramalhete, A. R. Amado, and J. C. Freitas. 2017. Inclusive housing program: The case of Oé-Cusse region in East Timor. *Frontier of Architectural Research* 6 (1): 74–88. https://doi.org/10.1016/j.foar.2016.12.001.

Apostol, L., and F. C. Mihai. 2012. Rural waste management: challenges and issues in Romania. *Present Environment and Sustainable Development* 6(2), 105–114. http://pesd.ro/articole/nr.6/2/10RWMCAIIR15102012105114.pdf

Badan Pusat Statistik Indonesia/BPS. 2018. Statistical Yearbook of Indonesia. Available in https://www.bps.go.id/id/publication/2018/07/03/5a963c1ea9b0fed6497d0845/statistik-indonesia-2018

Badan Pusat Statistik Indonesia/BPS. 2019. Statistical Yearbook of Indonesia. Available in https://www.bps.go.id/id/publication/2019/07/04/daac1ba18cae1e90706ee58a/statistik-indonesia-2019.html

Badan Pusat Statistik Indonesia/BPS. 2025. Statistical Yearbook of Indonesia. Available in https://www.bps.go.id/id/publication/2025/02/28/8cfe1a589ad3693396d3db9f/statistik-indonesia-2025.html

Baghel, A., and A. Thakkar. 2019. Bamboo: A resilient material for mass housing in earthquake prone zones of Gujarat. *Proceeding of Sustainable Built Environment*. https://www.researchgate.net/profile/Archana-Baghel/publication/332697132_Bamboo_A_Resilient_Material_for_Mass_Housing_in_Earthquake_Prone_Zones_of/links/5cc3f8a8a6fdcc1d49b22f15/Bamboo-A-Resilient-Material-for-Mass-Housing-in-Earthquake-Prone-Zones-of.pdf

Baldwin, C., T. Smith, and C. Jacobson. 2017. Love of the land: Social-ecological connectivity of rural landholders. *Journal of Rural Studies* 51: 37–52. https://doi.org/10.1016/j.jrurstud.2017.01.012.

Baye, K. 2017. The sustainable development goals cannot be achieved without improving maternal and child nutrition. *Journal of Public Health Policy* 38: 137–145. https://doi.org/10.1057/s41271-016-0043-y.

Blum, N. U., R. S. Wakeling, and T. S. Schmidt. 2013. Rural electrification through village grids—Assessing the cost competitiveness of isolated renewable energy technologies in Indonesia. *Renewable and Sustainable Energy Review* 22: 482–496. https://doi.org/10.1016/j.rser.2013.01.049.

Boliko, C. M., and D. S. Ialnazov. 2019. An assessment of rural electrification projects in Kenya using a sustainability framework. *Energy Policy* 133: 110928. https://doi.org/10.1016/j.enpol.2019.110928.

Canavan, C. R., L. Graybill, W. Fawzi, and J. Kinabo. 2016. The SDGs will require integrated agriculture, nutrition, and health at the community level. *Food and Nutrition Bulletin* 37 (1): 112–115. https://doi.org/10.1177/0379572115626617.

Celentano, G., E. Z. Escamilla, V. Göswein, and G. Habert. 2019. A matter of speed: The impact of material choice in post-disaster reconstruction. *International Journal of Disaster Risk Reduction* 34: 34–44. https://doi.org/10.1016/j.ijdrr.2018.10.026.

Cogut, A. 2016. *Open burning of waste: A global health disaster*. R20 Regions of climate action. https://regions20.org/wp-content/uploads/2016/08/OPEN-BURNING-OF-WASTE-A-GLOBAL-HEALTH-DISASTER_R20-Research-Paper_Final_29.05.2017.pdf

Dib, J. B., Z. Alamsyah, and M. Qaim. 2018. Land-use change and income inequality in rural Indonesia. *Forest Policy and Economics* 94: 55–66. https://doi.org/10.1016/j.forpol.2018.06.010.

Department of Economics and Social Affairs. 2019. *Accelerating SDG 7 Achievement: SDG 7 Policy Briefs in Support of the High-Level Political Forum 2019*. United Nations. https://sustainabledevelopment.un.org/content/documents/22877UN_FINAL_ONLINE_20190523.pdf

Díaz, J. P., and J. Romaní Fernández. 2016. An approximation to household overcrowding: Evidence from ecuador. *Revista Politécnica 37*(2), 1–8. https://revistapolitecnica.epn.edu.ec/ojs2/index.php/revista_politecnica2/article/view/678

Drahansky, M., M. Paridah, A. Moradbak, A. Mohamed, F. Owolabi, Abdulwahab Taiwo, M. Asniza, and S. H. Abdul Khalid. 2017. Introductory Chapter: Rural Waste Management Issues at Global Level 1–10.

Ezbakhe, F., R. Giné-Garriga, and A. Pérez-Foguet. 2019. Leaving no one behind: Evaluating access to water, sanitation and hygiene for vulnerable and marginalized groups. *Science of the Total Environment* 683: 537–546. https://doi.org/10.1016/j.scitotenv.2019.05.207.

Food and Agriculture Organization. 2018. *FAO'S work on family farming: Preparing for the decade of family farming (2019–2028) to achieve the SDGs*. https://www.fao.org/publications/card/en/c/CA1465EN/

Gan, X., J. Zuo, P. Wu, J. Wang, R. Chang, and T. Wen. 2017. How affordable housing becomes more sustainable? A stakeholder study. *Journal of Cleaner Production* 162: 427–437. https://doi.org/10.1016/j.jclepro.2017.06.048.

Gao, X., A. Xu, L. Liu, O. Deng, M. Zeng, J. Ling, and Y. Wei. 2017. Understanding rural housing abandonment in China's rapid urbanization. *Habitat International* 67: 13–21. https://doi.org/10.1016/j.habitatint.2017.06.009.

Gignoux, J., and M. Menéndez. 2016. Benefit in the wake of disaster: Long-run effects of earthquakes on welfare in rural Indonesia. *Journal of Development Economics* 118: 26–44. https://doi.org/10.1016/j.jdeveco.2015.08.004.

Giovannucci D., et al. 2012. Food and agriculture: The future of sustainability Study. *The Sustainable Development in the 21st century (SD21) Report for Rio+20*, New York: United Nations, 2012. https://doi.org/10.2139/ssrn.2054838

Gorbenkova, E., E. Shcherbina, and A. Belal. 2018. Rural areas: Critical drivers for sustainable development. *IFAC-Papers Online* 51 (30): 786–790. https://doi.org/10.1016/j.ifacol.2018.11.195.

Horton, S., and R. H. Steckel. 2013. Malnutrition: Global economic losses attributable to malnutrition 1900–2000 and projections to 2050. In *How much have global problems cost the earth? A scorecard from 1900 to 2050*, ed. B. Lomborg, 247–272. Cambridge: Cambridge University Press.

Howard, G., K. Charles, K. Pond, A. Brookshaw, R. Hossain, and J. Bartram. 2010. Securing 2020 vision for 2030: Climate change and ensuring resilience in water and sanitation services. *Journal of Water and Climate Change* 1 (1): 2–16. https://doi.org/10.2166/wcc.2010.105b.

Howard, G., J. Bartram, and World Health Organization. 2010. *Vision 2030: the resilience of water supply and sanitation in the face of climate change: technical report* (No. WHO/HSE/WSH/10.01). World Health Organization.

Jha, S. K., S. Mishra, B. Sinha, J. M. Alatalo, and R. Pandey. 2017. Rural development program in tribal region: A protocol for adaptation and addressing climate change vulnerability. *Journal of Rural Studies* 51: 151–157. https://doi.org/10.1016/j.jrurstud.2017.02.013.

Ji, X., J. Ren, and S. Ulgiati. 2019. Towards urban-rural sustainable cooperation: Models and policy implication. *Journal of Cleaner Production* 213: 892–898. https://doi.org/10.1016/j.jclepro.2018.12.097.

Khaleel, A., and I. Ngah. 2013. Understanding the sustainability of rural communities within a population density context: A case study of Shaqlawa, Kurdistan, Iraq. *Journal of Sustainability Science and Management* *8*(2), 254–267. https://www.cabdirect.org/cabdirect/abstract/20143079329

Kristiansen, S., and Pratikno. 2006. Decentralising education in Indonesia. *International Journal of Educational Development* 26 (5): 513–531. https://doi.org/10.1016/j.ijedudev.2005.12.003.

Lekic, O. Z., N. Gadzic, and A. Milovanovic. 2018. Sustainability of rural areas: Exploring values, challenges, and socio-cultural role. In: A. Fikfak, S. Kosanovic, and M. Konjar, editors. *Sustainability and Resilience Socio-Spatial Perspective*, 171–184. TU Delft Open. https://books.bk.tudelft.nl/press/catalog/book/isbn.9789463660303

Li, H. X., L. S. Ma, L. Lin, Z. X. Ni, X. R. Xu, H. H. Shi, and D. Rittschof. 2018. Microplastics in oysters *Saccostrea cucullata* along the Pearl River estuary, China. *Environmental Pollution* 236: 619–625. https://doi.org/10.1016/j.envpol.2018.01.083.

Li, Y., H. Westlund, and Y. Liu. 2019. Why some rural areas decline while some others not: An overview of rural evolution in the world. *Journal of Rural Studies* 68: 135–143. https://doi.org/10.1016/j.jrurstud.2019.03.003.

Litzow, E., S. K. Pattanayak, and T. Thinley. 2017. *Evaluating Rural Electrification: Illustrating Research Gaps with the Case of Bhutan.* Environment for Development Discussion Paper. http://hdl.handle.net/2077/66052

Liu, Z., S. Liu, H. Jin, and W. Qi. 2017. Rural population change in China: Spatial differences, driving forces and policy implications. *Journal of Rural Studies* 51: 189–197. https://doi.org/10.1016/j.jrurstud.2017.02.006.

Ma, W., G. Jiang, W. Li, T. Zhou, and R. Zhang. 2019. Multifunctionality assessment of the land use system in rural residential areas: Confronting land use supply with rural sustainability demand.

Journal of Environmental Management 231: 73–85. https://doi.org/10.1016/j.jenvman.2018.09.053.
Mihai, F. C., ed. 2017. *Solid waste management in rural areas*. London: InTech Open. https://doi.org/10.5772/66551.
Mihai, F. C., and M. J. Taherzadeh. 2017. Introductory chapter: Rural waste management issues at global level. In *Solid waste management in rural areas*, ed. F. Mihai. London: IntechOpen. https://doi.org/10.5772/66551.
Moharana, P. C. 2012. Rural solid waste management: Issues and action. *Kurukshetra, Journal on Rural Development* 60 (7): 30–34.
Mughal, A. W., J. Aldridge, and M. Monaghan. 2019. Perspectives of dropped-out children on their dropping out from public secondary schools in rural Pakistan. *International Journal of Educational Development* 66: 52–61. https://doi.org/10.1016/j.ijedudev.2019.02.004.
Øhlenschlæger, M., S. C. Christensen, H. Bregnhøj, and H. J. Albrechtsen. 2016. Submerged pond sand filter—A novel approach to rural water supply. *Water* 8 (6): 250. https://doi.org/10.3390/w8060250.
Omer, M. A., and T. Noguchi. 2020. A conceptual framework for understanding the contribution of building materials in the achievement of Sustainable Development Goals (SDGs). *Sustainable Cities and Society* 52: 101869. https://doi.org/10.1016/j.scs.2019.101869.
Pablo, D. J., and Javier, R. 2016. An approximation to Household Overcrowding: Evidence from Ecuador. *Revista Politécnica—Marzo 2016* 37 (2).
Parwanto, N. B., and T. Oyama. 2014. A statistical analysis and comparison of historical earthquake and tsunami disasters in Japan and Indonesia. *International Journal of Disaster Risk Reduction* 7: 122–141. https://doi.org/10.1016/j.ijdrr.2013.10.003.
Patunru, A. A. 2015. Access to safe drinking water and sanitation in Indonesia. *Asia and the Pacific Policy Studies* 2 (2): 234–244. https://doi.org/10.1002/app5.81.
Pemerintah Kabupaten Ende. 2015. *Peraturan bupati Ende nomor 19 tahun 2015 tentang rencana kerja pemerintah daerah (RKPD) Tahun 2016*. (*Ende regency regulation No. 19/2015 about regional government work plan for 2016*).
Perrucci, D. V., B. A. Vazquez, and C. B. Aktas. 2016. Sustainable temporary housing: Global trends and outlook. *Procedia Engineering* 145: 327–332. https://doi.org/10.1016/j.proeng.2016.04.082.
Prata, J. C., J. P. da Costa, I. Lopes, A. C. Duarte, and T. Rocha-Santos. 2020. Environmental exposure to microplastics: An overview on possible human health effects. *Science of the Total Environment* 702: 134455. https://doi.org/10.1016/j.scitotenv.2019.134455.
Reyna-Bensusan, N., D. C. Wilson, and S. R. Smith. 2018. Uncontrolled burning of solid waste by households in Mexico is a significant contributor to climate change in the country. *Environmental Research* 163: 280–288. https://doi.org/10.1016/j.envres.2018.01.042.
Rizal, M. F., and E. van Doorslaer. 2019. Explaining the fall of socioeconomic inequality in childhood stunting in Indonesia. *SSM-Population Health* 9: 100469. https://doi.org/10.1016/j.ssmph.2019.100469.
Rodić, L., and D. C. Wilson. 2017. Resolving governance issues to achieve priority sustainable development goals related to solid waste management in developing countries. *Sustainability* 9 (3): 404. https://doi.org/10.3390/su9030404.
Rodríguez-Pose, A., and D. Hardy. 2015. Addressing poverty and inequality in the rural economy from a global perspective. *Applied Geography* 61: 11–23. https://doi.org/10.1016/j.apgeog.2015.02.005.
Royuela, V., J. P. Díaz-Sánchez, and J. Romaní. 2019. Migration effects on living standards of the left behind. The case of overcrowding levels in Ecuadorian households. *Habitat International* 93: 102030. https://doi.org/10.1016/j.habitatint.2019.102030.
Rujner, H., G. Leonhardt, J. Marsalek, and M. Viklander. 2018. High-resolution modelling of the grass swale response to runoff inflows with Mike SHE. *Journal of Hydrology* 562: 411–422. https://doi.org/10.1016/j.jhydrol.2018.05.024.

Rustinsyah, R. 2019. The significance of social relations in rural development: A case study of a beef-cattle farmer group in Indonesia. *Journal of Co-Operative Organization and Management* 7 (2): 100088. https://doi.org/10.1016/j.jcom.2019.100088.

Sallu, S. M., C. Twyman, and L. C. Stringer. 2010. Resilient or vulnerable livelihoods? Assessing livelihood dynamics and trajectories in rural Botswana. *Ecology and Society*, *15*(4). https://www.jstor.org/stable/26268197

Sarmiento, Z. F., B. Sagala, and H. Siagian. 2019. The Sokoria geothermal system, Flores Island, Indonesia. *Geothermics* 82: 282–295. https://doi.org/10.1016/j.geothermics.2019.07.003.

Sen, S. M., and A. Kansal. 2019. Achieving water security in rural Indian Himalayas: A participatory account of challenges and potential solutions. *Journal of Environmental Management* 245: 398–408. https://doi.org/10.1016/j.jenvman.2019.05.132.

Sinki, K., Y. Fujii, and B. Shibazaki. 2021. Numerical simulations of the 1992 Flores tsunami using earthquake fault and landslide models. Synopsis of IISEE-GRIPS Master's Thesis. https://iisee.kenken.go.jp/syndb/data/20211209487648b2.pdf.

Straka, J., and M. Tuzová. 2016. Factors affecting development of rural areas in the Czech Republic: A literature review. *Procedia-Social and Behavioral Sciences* 220: 496–505. https://doi.org/10.1016/j.sbspro.2016.05.525.

Sudibyo, H., Y. S. Pradana, A. Budiman, and W. Budhijanto. 2017. Municipal solid waste management in Indonesia-A study about selection of proper solid waste reduction method in DI Yogyakarta Province. *Energy Procedia* 143: 494–499. https://doi.org/10.1016/j.egypro.2017.12.716.

Tadesse, T., A. Ruijs, and F. Hagos. 2008. Household waste disposal in Mekelle city, Northern Ethiopia. *Waste Management* 28 (10): 2003–2012. https://doi.org/10.1016/j.wasman.2007.08.015.

Torlesse, H., A. A. Cronin, S. K. Sebayang, and R. Nandy. 2016. Determinants of stunting in Indonesian children: Evidence from a cross-sectional survey indicate a prominent role for the water, sanitation and hygiene sector in stunting reduction. *BMC Public Health* 16 (1): 669. https://doi.org/10.1186/s12889-016-3339-8.

Ugochukwu, I. B., and M. I. B. Chioma. 2015. Local building materials: Affordable strategy for housing the urban poor in Nigeria. *Procedia Engineering* 118: 42–49. https://doi.org/10.1016/j.proeng.2015.08.402.

UNICEF. 2019. The State of The World's Children 2019—Children, food and nutrition: Growing well in a changing world. https://www.unicef.org/media/60806/file/SOWC-2019.pdf

United Nations. 2015. *Transforming our world: the 2030 Agenda for Sustainable Development*. https://sdgs.un.org/2030agenda

United Nations Environment Programme. 2015. *Global Waste Management Outlook*. https://www.unep.org/resources/report/global-waste-management-outlook

United Nations Development Programme (UNDP) and United Nations Research Institute for Social Development (UNRISD). 2017. *Challenges and Opportunities in the Implementation of the Sustainable Development Goals*. https://www.undp.org/serbia/publications/challenges-and-opportunities-implementation-sustainable-development-goals

Walkinshaw, C., P. K. Lindeque, R. Thompson, T. Tolhurst, and M. Cole. 2020. Microplastics and seafood: Lower trophic organisms at highest risk of contamination. *Ecotoxicology and Environmental Safety* 190: 110066. https://doi.org/10.1016/j.ecoenv.2019.110066.

Wang, F., Z. Cheng, A. Reisner, and Y. Liu. 2018a. Compliance with household solid waste management in rural villages in developing countries. *Journal of Cleaner Production* 202: 293–298. https://doi.org/10.1016/j.jclepro.2018.08.135.

Wang, J., S. D. Maier, R. Horn, R. Holländer, and R. Aschemann. 2018b. Development of an ex-ante sustainability assessment methodology for municipal solid waste management innovations. *Sustainability* 10 (9): 3208. https://doi.org/10.3390/su10093208.

WHO and UNICEF. 2017. *Progress on household drinking water, sanitation and hygiene 2000–2017*. https://apps.who.int/iris/handle/10665/329370

Chapter 9
Eating as Social Practice: Study on Stunting Children's Family Eating Pattern in Nangapanda

Raphaella Dewantari Dwianto, Putri Ayu Iramaya, Grace Wangge, and Dini Suciyanti

Abstract This chapter focuses on eating as a social practice of families in urbanized rural areas where one in three children are stunting. The argument is based on a survey conducted in October 2019 on families of young children (age 1–9 years old) in Nangapanda subdistrict, Ende District of East Nusa Tenggara Province. Despite the successful efforts from various institutions to reduce stunting in the area, Statistics of Nangapanda subdistrict shows one in three children experience stunting, the figure is further confirmed in the 2009 survey. Standing on the duality of agency-structure from theory of structuration, the everyday lived experience of eating in the family leads to understanding of eating as a social practice, which can explain the prevailing of stunting. While studies on stunting in Indonesia have focused on nutrition, parenting, and eating behavior, and though these studies contribute to both science and the policy-making process, study in this chapter contributes to further understanding of prevailing stunting in the area with findings show not only knowledge of mothers as respondents of the 2009 survey who are the decision makers on family eating habits, but also how the respondents answer the challenges in providing healthy food for the family.

Keywords Duality of structure-agency · Eating · Social practice · Stunting

R. D. Dwianto (✉) · P. A. Iramaya
Department of Sociology, Faculty of Social and Political Sciences, Universitas Indonesia, Depok, Indonesia
e-mail: raphaella.dwianto@ui.ac.id

P. A. Iramaya
e-mail: iramaya.putriayu@gmail.com

P. A. Iramaya
Jakarta, Indonesia

G. Wangge · D. Suciyanti
Southeast Asia Ministers of Education Organization–Regional Center for Food and Nutrition (SEAMEO/RECFON)/Pusat Kajian Gizi Regional, Universitas Indonesia, Depok, Indonesia
e-mail: gwangge@gmail.com

M. Budianta et al. (eds.), *A Liveable Kampung*, Engaging Indonesia,
https://doi.org/10.1007/978-981-96-7866-2_9

9.1 Introduction

Indonesia is one of the countries with a high prevalence of stunting. Data from WHO in 2018 reveals that in Indonesia 36.4% of under 5 years old infants are stunting, putting the country in the third place with the highest stunting prevalence in the South-East Asian Region. Among 34 provinces in Indonesia, East Nusa Tenggara province has the highest prevalence of stunting, with 43.82% of the infants under 5 years old suffering from stunting. Stunting is a condition of failure to grow in children (infants under 5 years old) resulting from chronic malnutrition. Malnutrition is caused by multi-dimensional factors. Among others is poor parenting practices, including mothers and pregnant women's lack of knowledge on health and nutrition. Another cause is lack of household or family access to nutritious food, due to poverty. Limited access to nutritious food in Indonesia contributes to anemia experienced by 1 in 3 pregnant women.

Majority of health policies on stunting in Indonesia focus on the importance of increasing access and changing behavior of individuals (especially mothers and under-five children) on nutritious food intake. Focusing only on individual nutritional intake and eating behavior undermine the significance of social context of the local society. The authors argue that eating—including eating nutritious food—is not just a behavior, but it is a social practice. Understanding eating as a social practice should be included in the discussion of stunting in Indonesia and in stunting reduction and prevention programs.

Research for this chapter focuses on eating as a social practice in the community in Nangapanda subdistrict, Ende District of East Nusa Tenggara Province. Despite the effort of the central and local government in eliminating and preventing stunting in the region, the decline in prevalence rates has not succeeded in placing this region as a stunting-free area. The author proposes to include observation and analysis on underlying social relations and sociocultural context relating to eating nutritious food, to have a comprehensive understanding on the issue. Studies that focus on eating as a social practice in relation to stunting, particularly in Indonesia, are still very limited. It is therefore important to understand the social context with its underlying social relations on eating nutritious food. The findings could contribute to the decision-making process regarding policies on stunting prevention in Nangapanda.

9.2 Conceptual Framework

Studies on stunting in Indonesia have focused on nutrition, parenting, parenting nutrition, and eating behavior (Abouda et al. 2013; Lestari et al. 2014; Maywita 2018; Nugraha et al. 2019; Achadi et al. 2020; Flynn et al. 2020). These studies have significantly contributed to both science and the policy-making process. This chapter focuses on the problem of stunting in Indonesia, especially in the province of East Nusa Tenggara by trying to understand the social context of the local community, which influences the health behavior of the society in question. In other words, the study on stunting in this chapter tries to use a sociological perspective, which argues that social relationships underlie eating patterns and collective eating, starting from the unit of families. Understanding these social relationships means analyzing the sociocultural context to understand eating behavior in the society where one third of children are confirmed as stunted.

The conceptual framework for studying the relationship between social conditions and the eating behavior of the population has not been widely developed. This chapter borrows concepts from Delormier et al. (2009) who use Giddens' structuration theory (1984, 1993). The structuration theory applied by Delomier et al. specifically raised the concepts of social practice, the duality of social structure, and agency. Through these three concepts, eating is not only understood as individual choice, but more as a social practice, where duality of social structure and agency works. Giddens theory of structuration emphasis on an interplay of 'agency' and 'social structure' (rules and resources), in the study of food choice patterns. Based on theoretical framework of theory of structuration and following footsteps of Delormier et al. (2009), this chapter aims at conceptualizing eating as social practice, to get insights on how the social structure as rules and resources creates condition of practice, in which social structure does not merely determine individual action, but it is enacted and reified by people through choices they make during social practices (Giddens 1984).

Related to theory of structuration, concepts of social context, social practice, social structure, agency, and family will be described in the following paragraphs. The social context is 'the local configuration of social relations which are composed of social structures such as class, race, and gender; institutional practices, collective and individual behavior, and intersecting personal biographies' (Poland et al. 2006: 60 as quoted in Delormier et al. 2009).

Giddens defined social practice as the situated activities of social actors which happen in the flow of daily life or in context, integrating the concepts of agency and social structure. Social practices are skillful procedures, methods, or techniques that are appropriately performed by social agents (Giddens 1984). Agency reflects the capacity for actors to intervene in the world and to influence a specific process or state of affairs (Giddens 1984). Social structure according to Giddens can be distinguished as follows. First, as two aspects of rules, that are codes of conduct (normative elements) and codes of signification (meaning). Secondly as two kinds of resources: allocative and authoritative. Social structures comprising two aspects of rules and two kinds of resources are linked to social interaction via modalities which are interpretive schemes, norms, and facilities. Structures of significance, domination, and legitimation are also important (Delormier et al. 2009).

Giddens theory of structuration conceptualizes agency (voluntary action) and structure (rules and resources) as a 'duality' not dualism, where both exist interdependently in a reciprocal relationship. In 'duality of structure', structures act as the medium of social practice and as the outcome of social practice (1993).

Regarding family, in this chapter family is defined more as the domain of social practices, rather than family as unit analysis nor as a structure with a certain position for certain people. Eating in the family is considered a lived experience that is taking place in a social group, i.e., family. Eating in this case includes obtaining food, preparing, consuming food, and food choice. As the domain for lived experience, the examination of daily food-related activities, i.e., food procurement for the family, food preparation, meal cooking are key empirical basis for understanding eating as social practice. By examining eating in the family as social practice, there will be a conceptual entry point for apprehending the underlying social relations which connect people in the social world, and which generate population eating patterns (Delormier et al. 2009).

9.3 Methods and Area Characteristics

9.3.1 Methods

East Nusa Tenggara province was chosen for this study due to the province's high (highest) prevalence of stunting in Indonesia. The study focused on Nangapanda subdistrict (*kecamatan*), Ende district/regency (*kabupaten*). Three villages of the Nangapanda subdistrict, namely Ndoturea, Ndororea, Ndororea 1, are selected. Data collections were conducted during 2019. Samples were collected from 1000 children under five years old and primary school-age children, to measure the level of nutrition. Then 200 parents of the children were randomly selected to get an overview of eating practices in the family. Prior to description of the findings, the following part will be the description of the area and its demographic characteristics.

9.3.2 Area and Demographic Characteristics

The study was conducted in September–October 2019, covering Nangapanda subdistrict (*kecamatan*) in Ende district/regency (*kabupaten*) of East Nusa Tenggara Province in Flores Island, Indonesia. Nangapanda subdistrict is a combination of coastal area and mountain area, divided into 85 villages (based on Nangapanda subdistrict statistics of 2017). This study covers 24 villages: Kekandere, Ndeturea, Ndorurea 1, Ndorurea, Rhaporendu, Bheramari, Nggorea, Anaraja, Ondorea, Ndetuzea, Embuzozo, Jhengarengga, Nagakeo, Oja, Rapowawo, Rendurua, Sanggaroro, Tenari, Tanazozo, Tendarea, Timbazia, Tiwerea, and Zozoziea. In the year 2019 when the study was conducted, the population of Nagapanda sub district was 22,859 people, with density of 107 people/km2, and ratio of male to female of 98. In the earlier years of 2000s Nangapanda sub district showed an increase year per year in its population between 2 and 5% of (see Table 9.1). However, in 2019 the statistical report pointed to a −0.3% in population growth. Regarding local administrative associations of neighborhood associations (*Rukun Warga*/RW) and neighboring household groups (*Rukun Tetangga*/RT), in the year 2017 there were 147 neighborhood associations (RW) with 268 neighboring household groups (RT) (See Table 9.2, data for 2019 is not available), with around 5700 households (see Table 9.3). According to the Bureau of Statistics of Ende district, Nangapanda subdistrict is the second most populous subdistrict (see Table 9.1) (Fig. 9.1).

In the area of study, small-scale plantation and farming is the main livelihood in the mountain area. While in the coastal area, the source of income ranges from fishing, bluestone mining and weaving. Some individuals also engage in jobs at government offices or private sector. The area of study also includes people migrating to other provinces of island in Indonesia, such as Kalimantan/Borneo Island to find work, or even migrating to neighboring country Malaysia, often as illegal migrant workers. Regarding ethnicity of people in Nangapanda sub district though the majority of the population belongs to Ende ethnic group, other ethnic groups like Bajawa, Jawa, Lio, Manggarai, Ngada, Sika, Soe, Sumba, Nagakeo, Ngada, and Ria, also live in the area although not in significant numbers.

Table 9.1 Population growth in Nangapanda subdistrict

	2009	2011	2013	2015	2017	2019
Population	19,756	19,895	20,092	23,294	24,192	22,859
Density (per km^2)	93	93	99	109	116	107
Population growth (%)	2.79	2	5	0	2.7	−0.3
Ratio male–female	92	91	89	95	95	98

Table 9.2 Government local administrative associations

	2009	2011	2013	2015	2017	2019
Area	NA	3	3	NA	NA	NA
Village	60	57	56	85	85	NA
Neighborhood association (RW)	106	106	107	147	147	NA
Neighborhood household (RT)	108	198	197	268	268	258

Table 9.3 Number of households

	2008	2009	2010	2011	2014	2016
Number of households	2165	802	2992	4711	5473	5773

Fig. 9.1 Nangapanda subdistrict on map of Indonesia (*Source* Badan Informasi Geospasial)

9.4 Findings and Discussion

9.4.1 Respondents' Demographic Characteristics

Following findings from samples collected from 1000 children under five years old and primary school-age children, one in three children are confirmed as stunted. Concerning profile of children for the survey, in terms of gender, 48% are female and 52% were male. The age of the children range between one year old to 9 years old with the majority in the age of 7 years old and 8 years old. Seven years old children cover 69.9% and 8-year-old children cover 25% of the total number of children surveyed. In other words, the majority of children surveyed are in their first year or second year of primary school (Table 9.4).

Table 9.4 Children's gender and age

Gender	%		Age	%
male	51.5		1 years old	1.5
female	48.0		2 years old	0.5
			6 years old	2.5
			7 years old	68.5
			8 years old	24.5
			9 years old	0.5

Table 9.5 Respondents and spouse's age

Age	Respondent (%)	Spouse (%)
20–29 years old	9.0	1.5
30–39 years old	49.0	41.0
40–49 years old	35.0	38.0
50–59 years old	2.5	11.0
60–69 years old	0.5	0.5

Respondents for a survey on eating as social practice are adults who are responsible for the children, the majority were the mothers (the terminology of respondent and mother will be used interchangeably). Respondents' age (mothers or grandmothers) range between 20 and 69 years old, with the majority in their 30 s (49%) and 40 s (35%). Nine percent of the respondents are in their 20 s and 2.5% are in their 50 s.

Age of respondents' spouses in the survey also shows a similar trend, in which most of the spouses (husbands) are in their 30 s (41%) and 40 s (38%). Spouses in their twenties covers 1.5% and spouses who are older than 50 years old covers 11.5%. Looking at the age of the majority of the respondents and their spouse it can be said that respondents are in their productive age (Table 9.5).

About educational background (see Table 9.6), 32.5% of respondents finished primary school, 21.5% graduated from junior high school, and 23.5% finished senior high school. 5.5% finished higher education. Yet 11.5% did not finish primary school and 3.5% did not even go to school at all. Regarding their spouse, the husbands' educational background echoes those of the respondents (the wives). Respondents' spouses who did not get any education and who did not finish primary education reached 15% of total respondents. The majority finished their primary education (32.5%). About 21.5% finished their junior high school and 23.5% finished their senior high school. Comparing educational background between respondents and their spouse, respondents who didn't go to school at all cover only half of the spouse who didn't get to school at all; no schooling for respondents is 1.5% whereas no schooling for spouse covers 3.5%. Among the respondents it seems that between primary school and Junior High School there are more men finishing their schooling compared to women. However, when it comes to finishing Senior High School there are more women who finished their High School compared to men. This also echoes in the higher education level where there are more women finishing their higher education compared to men among the respondents. This trend is in line with statistical data of East Nusa Tenggara province, where in 2019 the average total number of years in school was 7.8 years.

On the occupation of respondents, as summarized in Table 9.7, 53% identify themselves as housewives. Next to the number of housewives among respondents, come those who work as farmers in their own land or work for landowners (combined with fisherman and stock farmer), covering 35% of the total respondents. Those who work as small-scale traders and self-employed made up 3% of the total respondents. Those who are employees of government offices and private companies cover 4.5%.

Table 9.6 Respondents and spouse's education

Education	Respondent (%)	Spouse (%)
No schooling	1.5	3.5
Primary < 3 years	10.0	11.5
Finished primary	41.0	32.5
Finished junior high	14.0	21.5
Finished senior high	25.5	23.5
Finished higher education	7.5	5.5

Table 9.7 Respondents and spouse's occupation

Occupation	Respondent (%)	Spouse (%)
Housewife/housework	53.0	3.0
Land-owner farmer	30.0	56.0
Farm workers	4.0	5.0
Private employees	2.5	5.0
Government employees	2.0	3.0
Small-scale traders	1.5	2.5
Self-employed	1.5	7.0
Fisherman	0.5	5.5
Stock farmer	0.5	0.0
Driver	0.5	2.0
Laborer	0.5	2.0
Unemployed	0.5	0.0
Migrant worker	0.0	3.0
Others	3.0	4.0

The occupation of their spouses (husband) is mostly in farming. Spouses working as farmers in their own land, or working for landowners, combined with stock farmer and fisherman covers 66.5%. Compared to respondents, there are more spouses working as small-scale traders and self-employed (both combined as 9.5%, compared to 3% of the respondents). And slightly more spouses work as employees in government offices or private companies (8–4.5%). Among spouses, there are those who also worked as migrant workers in other parts of Indonesia, or outside of the country.

With the majority working as farmers, the number of people in the same house also shows that most respondents and their spouse live in a house of a big family in terms of number of people under the same roof, which also shows that in the area extended family prevails. Two third of the respondents (64.5%) live in a house where there are 4–6 people. A quarter of the respondents (25%) live in a house with 7–9 people in the house. Data from the survey also shows respondents living together with 10–12 people under the same roof, or even 13–15 people (the total covers 3%).

Table 9.8 Household number of person and monthly income

Number of person in the household	%	Household monthly income	%
1–3 person	7.0	0–Rp 500,000	65.5
4–6 person	64.5	Rp 500,001–Rp 1,000,000	19.5
7–9 person	25.5	Rp 1,000,001–Rp 1,500,000	5.5
10–12 person	2.0	Rp 1,500,001–Rp 2,000,000	2.0
13–15 person	1.0	Rp 2,000,001–Rp 2,500,000	2.5
		Rp 2,500,001–Rp 3,000,000	1.5
		Rp 3,000,001–Rp 3,500,000	0.5
		Rp 4,500,001–Rp 5,000,000	2.5

Regarding monthly income of respondent's household. two third calculated their monthly income as maximum of Rp 500,000 rupiah (or approximately USD 35 based on exchange rate in 2019), and one-fifth has monthly income between Rp 500,000 and Rp 1,000,000 (or approximately USD 35–70). Only 7.5% has monthly income between Rp 1,000,000 to Rp 2,000,000 (approximately USD 70–140). Four percent live on a monthly income between USD 140 and 210. Compared to Indonesian average monthly income which is Rp 4,700,000 (approximately USD 326), respondents' household monthly income is much lower (Table 9.8).

The large number of family members under one roof, coupled with a monthly income that is far below the average monthly household income in Indonesia, clearly shows the underprivileged economic condition of the respondents and posed challenges to prevention and treatment of stunting in children. Under such family and economic conditions, when it comes to food and meal for the family, respondents show their way of coping with the conditions.

9.4.2 Eating as Structure of Family Affairs

For Indonesians in general, eating has always been an important aspect of life, not because it is an essential for continuation of one's life, but more of the social meaning in the practice called eating. Eating with other people in a social gathering is as important as eating to keep one alive. Respondents in the survey also do the same thing. In this part, the meaning of eating together with family at home as their lived experience that is taking place in a social group will be the focus of discussion. Eating in this case includes obtaining food, preparing, consuming food, and food choice.

Among respondents in Nangapanda subdistrict, eating at home is considered one of family affairs, as a habit that is taken for granted since it is always there, or, borrowing the term from Giddens, it is the 'rule' (structure) to have meals at home together. Every day, members of the family manage to spend time together at home. As Fig. 9.2 describes, dinner time is when families gather to eat together every day

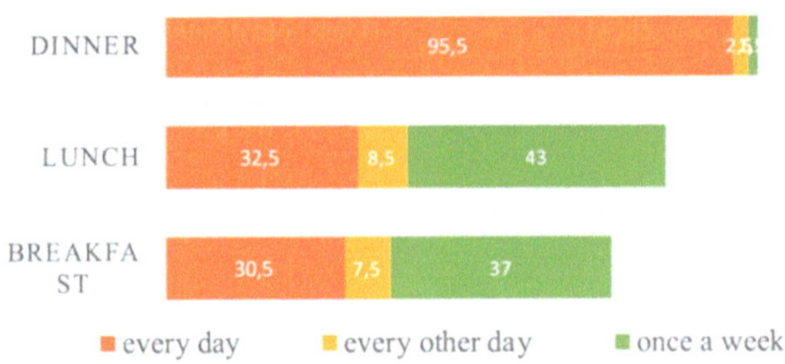

Fig. 9.2 Frequency of family eating together

(95.5%). Not only at dinner do families eat together. One third of respondents answer that they eat together at breakfast and lunch every day. Other responses show that at least once a week those who cannot eat breakfast and lunch every day at home with other members of family, manage to spare time to have these two meals together. Looking once again at the occupation of the respondents (Table 9.7), half of the respondents work outside their home, as farmer, small-scale trader, or employee. It is in line with which meal they can always share with other family members at home. In the morning, these respondents leave their house to go to their workplaces, be it the farm or plantation, or offices. These workers spend the whole day at their workplaces, and return home in the afternoon or evening, in time to prepare dinner and have the meal together with other members of the family.

Whenever family members gather to have a meal together at home, although two-thirds eat their meal probably in silence, there is a 38.5% who eat while doing other things. Among those other things that they do while having a meal at home is chatting with each other (24.7%). This affirms that eating at home with other family members is one form of social gathering, where an individual has a chance to have communication and share experiences and thoughts with other members. Other activities during meals also include watching TV (14.3%). Among respondents in Nangapanda subdistrict, television functions as a source of reliable information (Fig. 9.3). This is in contrast with their counterparts in urban areas, especially urban areas in Java Island, where television is starting to lose its role as the main source of information as people rely more on social media for information. The results of a survey by the Ministry of Information in 2017 showed that for the areas of Bali and Nusa Tenggara, hand phone ownership reached 45.24%, lower than the average in Indonesia, which was 66.31%, and far below the island of Java which reached 86.6%. However, although the number of ownerships of mobile phones is still low, some of the respondents answered that while eating together at home, other activities they do are playing with their mobile phones.

During family meals, the room in the house where the family has their meal together is the kitchen or living room. Among respondents, 52% have their family eat together in their kitchen. As more than half of respondents have their family meal in their kitchen, it can be said that for families in Nangapanda subdistrict, the kitchen

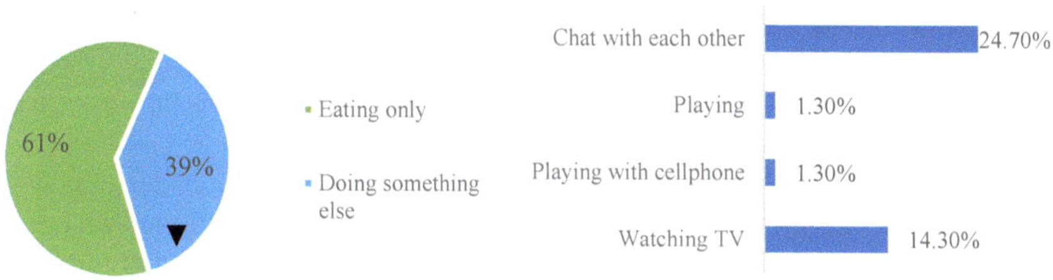

Fig. 9.3 Activity while family eating together at home

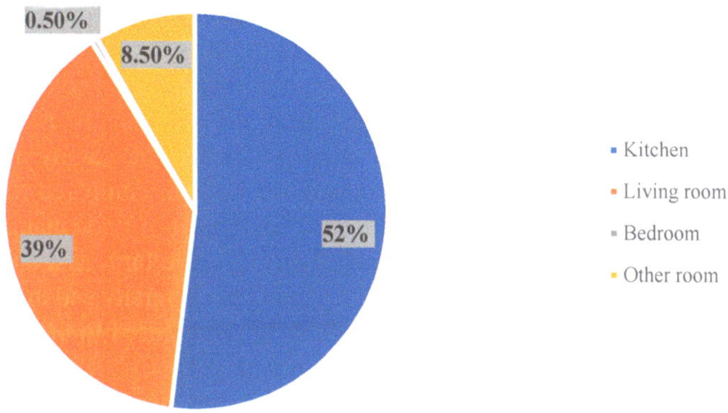

Fig. 9.4 Room for family eating together

is multifunctional and serves as the social space for the family when it comes to eating together. Kitchen is not a mere culinary space, but a shared space, or the 'heart of the home'. Second in place is the living room. Thirty nine percent of respondents have their family meal in the living room (Fig. 9.4).

In Nangapanda subdistrict, where the family is generally a large family or extended family living under one roof, the habit of eating together in the family, and making mealtimes an opportunity to share thoughts to each other, then makes the kitchen not only a place to cook food but also an arena for socializing with extended family members, all these form the structure that each member of the extended family follows, including the women who were respondents for this survey. Yet, this structure is not a rigid one, but always negotiable, by the respondents as actors in their everyday life activities concerning food and eating for their family.

9.4.3 Food on the Table: Actor's Agency on Family Meal

As members of families in Nangapanda subdistrict, the social structure consisting of rules on family meal, and education/economic institutions which give challenges to the respondents, are the everyday life facts for the respondents. In this part, findings on respondents as actors who are in the decision-making position concerning healthy food for the big family will be described in detail, to understand how they exert their agency in the above-mentioned structure.

To begin with, the majority of respondents understand the importance of putting healthy food on their family table. Not only do the respondents understand what food is considered healthy, they also put effort in balancing the composition of food they prepare for the family. In the composition of food, 41.8% consist of protein, 38.1% are vegetables, 16.1% is carbohydrate. Respondents can have easy access to food sources in the form of protein and vegetables, because the Nangapanda subdistrict consists of mountainous areas that produce vegetables, and there are also coastal areas with marine products. They also put some fruits on their family table, though the percentage is much lower (3.7%) compared to protein, vegetables, and carbohydrates. The thing that needs special attention is that, from the respondents' answers, there is no statement that they include milk or dairy products in the family meal. It can be said that access to milk and dairy products are limited, compared to access to protein, vegetables, and carbohydrates. In 2020 the Regent of Ende inaugurated the establishment of the Nangapanda Livestock Industry cattle farm as a collaborative effort of the Ende Regency Government with a national university and investors. The geographical proximity to milk and dairy products can be used to open the access for the residents in Nangapanda subdistrict (Fig. 9.5).

For carbohydrates, 88% of respondents choose to serve rice as the main source of carbohydrate for the family. Nangapanda is one area in Ende regency with wetland paddy fields. Other than rice, the choice includes sweet potato, corn—also in the form of corn rice—and cassava. The respondents understand the importance of eating carbohydrate and put it on the family table (Figs. 9.6, 9.7).

Fig. 9.5 Composition of food

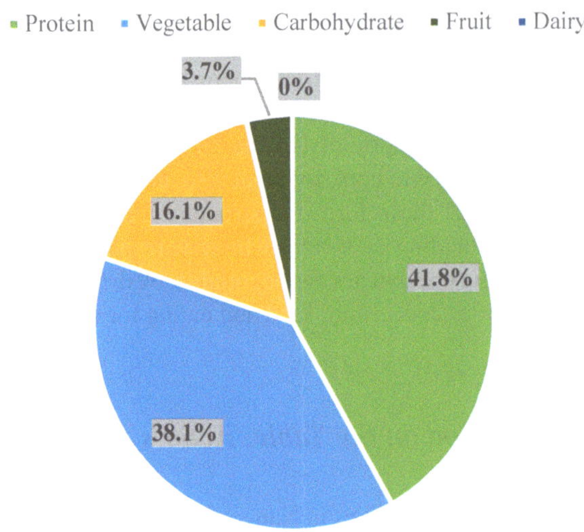

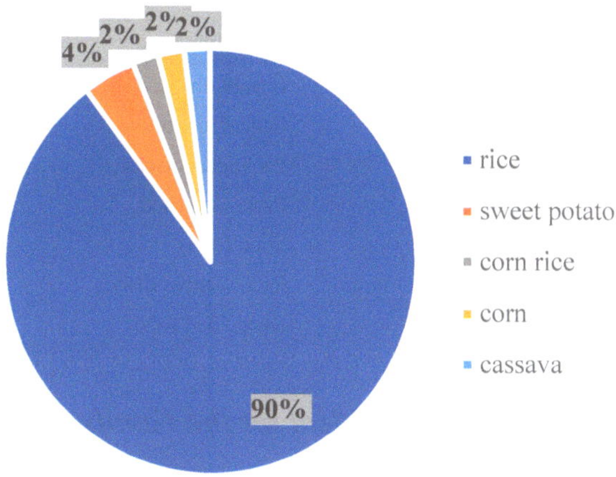

Fig. 9.6 Variety of carbohydrate

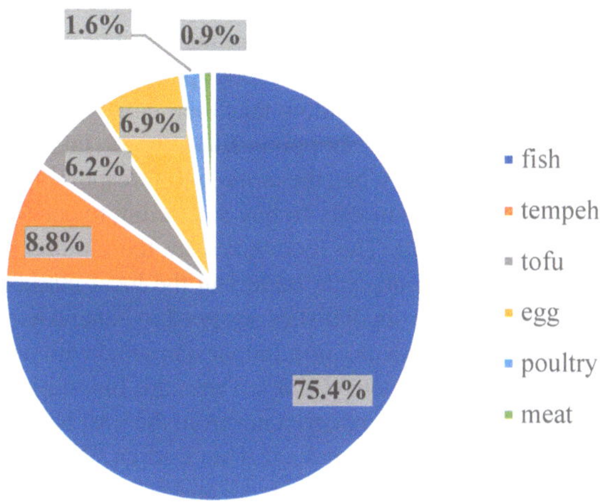

Fig. 9.7 Variety of protein

The main source of protein for families in Nangapanda is fish (75.4% of total protein). Animal protein served on the table includes egg (6.9%) and poultry (1.6%). Plant protein includes tempeh (8.8%) and tofu (6.2%).

Looking at the food composition for families in Nangapanda, it can be said that family members in Nangapanda eat balanced and healthy food. It seems that they also do not consume unhealthy food, such as instant noodles, packaged snacks, processed food, and deep-fried food (see Table 9.9). In the following paragraphs, the decisions made by respondents in the process of food preparation will be described in detail.

Table 9.9 Respondent's knowledge on unhealthy food

Unhealthy food	%
Instant noodle	80
Others, including	20
Packaged snacks	
Processed food	
Stale food	
Rice without dishes	
Deep fried food	

9.4.4 Preparing Meal for the Family

With sufficient knowledge on healthy food, respondents are also in the position of deciding the menu, doing the shopping, and cooking for the family (see Table 9.10). When buying ingredients for cooking, most of the respondents (92%) choose to buy fresh ingredients, and never choose processed food (68.5%).

Most respondents seldom take their children to the local market when they have to do some groceries shopping. But some do take their children to the local market from time to time. When the mothers (the respondents) are doing the shopping, some children ask their mother to buy something to eat, other than those in the shopping list of the mother. The food asked by the children range from Indonesian dishes often sold by small-scale traders in the local market, such as meat ball soup (*mie bakso*), rice with small dishes wrapped in banana leaves (*nasi bungkus*), mung bean porridge (*bubur kacang ijo*), Indonesian salad with peanut dressing (*gado-gado*), and sweet potato compote (*kolak*). These are Indonesian dishes that are not originally from Nangapanda subdistrict nor from the East Nusa Tenggara province. Apart from Indonesian dishes, other food that children often ask their mother to buy includes cakes, cookies, doughnuts, bread, and packaged snacks. Some children also ask for ice cream and milk. Regarding these dairy products, it is contrary to the data on composition of food, where no respondents serve milk as part of a family meal (Tables 9.11, 9.12, 9.13).

After doing groceries shopping, these mothers are also responsible for cooking the meal. In planning menus and cooking meals, respondents rarely try new menus (Table 9.10). They follow a menu that has been passed down from generation to generation in their respective families. They follow family recipes. Since family recipes are orally thought, respondents do not need to refer to written cooking recipes

Table 9.10 Preparing a family meal

	Planning menu	Grocery shopping	Cooking	Following family recipe	Trying new menu
Yes	93.5	90.5	94.5	86	11.5
No	6.5	9.5	5.5	13	85.5

Table 9.11 Respondent's choice of ingredients

	Fresh ingredients	Processed food
Never	3.0	68.5
Seldom	1.5	13.5
Sometimes	3.5	14.5
Often	92.0	3.5

Table 9.12 Taking children to buy groceries in local market

	Taking children to market	Children asking to buy food at
Never	70.5	61.5
Seldom	15.5	12.0
Sometimes	8.0	6.5
Often	6.0	5.5

Table 9.13 Food asked by the respondent's children in the local market

Indonesian local dishes	%	Non-Indonesian	%
Meat ball soup (*mi bakso*)	9.4	Cakes	15.1
Rice + small dishes (*nasi bungkus*)	5.7	Cookies	11.3
Mung bean porridge (*bubur kacang ijo*)	3.8	Doughnut	1.9
Indonesian salad (*gado-gado*)	1.9	Bread	1.9
Sweet potato compote (*kolak*)	1.9		
Fried snacks	%	Other	%
Fried peanuts	3.8	Jelly	1.9
Crackers (*kerupuk*)	1.9	Fruits	3.8
Other snacks (including packaged)	22.6		
		Dairy product	%
		Ice cream	9.4
		Milk	3.8

(see Table 9.14). Seventy four percent never need a recipe when cooking, whether it is a cooking recipe from a cookbook, or from magazine or television. They do not use the internet either for getting ideas of a menu nor for finding cooking recipes.

Cooking food that has been on the table of the family for generations is the natural thing for respondents. This confirms the importance of family for the people in Nangapanda subdistrict. When it comes to getting information on healthy food, asking mother, grandmother, and parents is the second highest choice for the respondents. They understand that they need to know more about healthy food for the family. The first source of information they chose to rely on is government officials (44.5%), including schoolteachers and medical staff at community health centers (*Pusat Kesehatan Masyarakat*). The second source are older family members including mothers,

Table 9.14 Cooking and recipe

Cooking recipe	Need recipe when cooking	Remember recipe by heart	From cookbook	From magazine	From TV	From the internet
Never	74.0	69.5	83.5	84.0	82.5	82.0
Seldom	6.0	2.0	1.5	1.0	1.5	1.0
Sometimes	5.0	1.0	NA	NA	1.5	1.0
Often	14.5	13.5	NA	NA	NA	1.0

grandmothers, and parents (19.5%). In the third place is relatives, friends, neighbors (15%). In fourth place is television (13%). Worth noting is the fifth place, which is the internet. Although only 3.5% or respondents try to get information on healthy food from the internet, the fact that respondents as actors begin to rely on sources that they do not know personally shows a shift among the people in Nangapanda (Fig. 9.8).

When the information they seek is not limited to healthy food, but more about general information, the trend is similar with information on healthy food, in which they rely on government officials. Yet, the second reliable source of information on general matters is not the elderly in the big family (mother, grandmother, parents) but television. Older members in the big family, relatives—friends—neighbors come in the third and fourth place. The Internet—again—begins to gain trust as a source of reliable information (3.5%). Regarding television, although it begins to lose its function as the main source of information for the people in Nangapanda, 45% of respondents still watch television every day. It can be said that television still plays

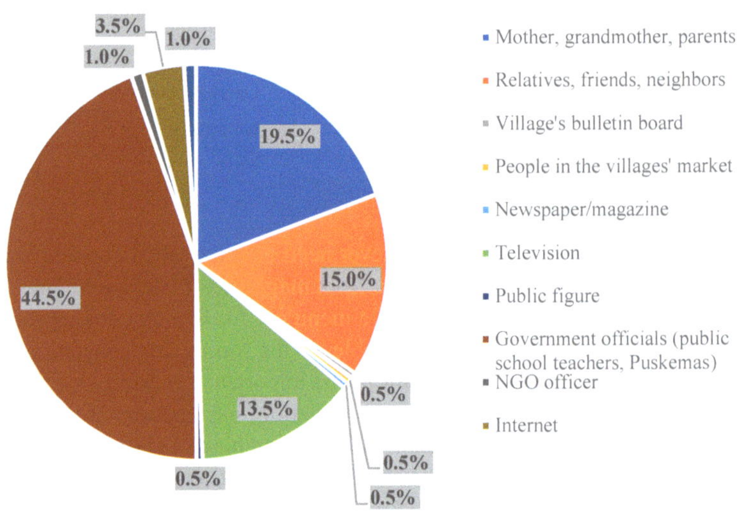

Fig. 9.8 Main source information on healthy food

an important role as a medium of information in Nangapanda, but its position begins to weaken, and might be replaced by the internet (Figs. 9.9, 9.10).

The survey conducted in Nangapanda subdistrict involving respondents of women (mothers) in families with young children shows that in the flow of daily life these mothers as social actors or social agents conceptualize healthy food, planning the menu for the family, doing the groceries, cooking the meal and putting them on the family table, by following skillfully the habits and rules of local big families. From the perspective of Giddens (1984), the procedures, methods, and techniques regarding healthy food for the family are appropriately performed by these mothers, even though their average years in education are slightly below Indonesian women in general, and their household monthly income falls into the poor household category.

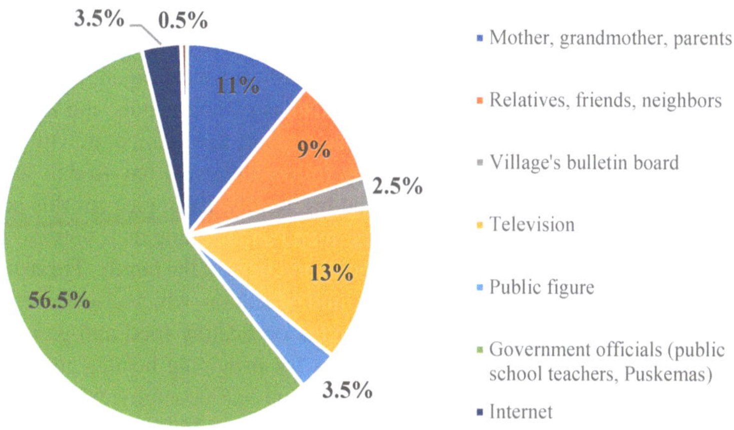

Fig. 9.9 Main source information on various matters

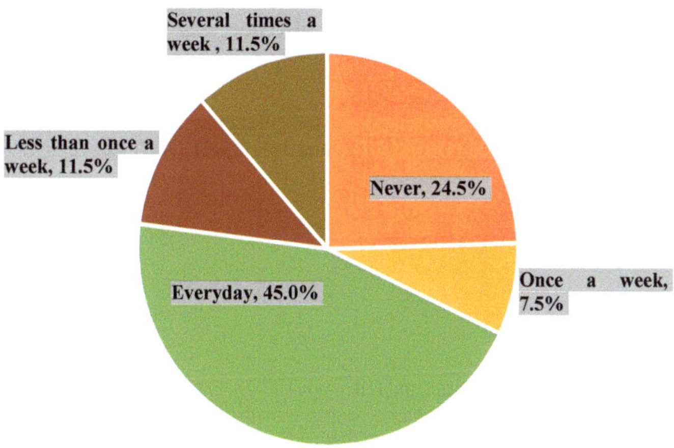

Fig. 9.10 Use of TV as a source of information

9.5 Conclusion

Understanding how the social context affects health behavior is necessary for more powerful interventions. Eating is embedded in the flow of day-to-day life. People's eating patterns form in relation to other people, alongside everyday activities that take place in family groups (Delormier et al. 2009). The challenges of eliminating stunting completely from the province of East Nusa Tenggara lie not only in 'post-swallowing' nutritious food which are in the domain of research in biochemistry, biology, pathology, and physiology, but also in the 'pre-swallowing' domain of society, culture, experience, and behavior (following Crotty 1993 as quoted in Delormier et al. 2009). It is therefore necessary to understand eating not only as an individual choice, but more as a social practice, where duality of social structure and agency works.

Mothers as respondents of the 2009 Nangapanda survey are the decision makers for their family meal. In everyday life, they work out their 'agency' not only as providers of healthy food for the family but also as planner, cook, and information seeker. While doing these activities, they also support family income with working in the house or outside of the house. They follow the usual day-to-day 'codes of conduct' and 'codes of signification' of the big family. Examples for this are their preserving their mothers and grandmothers' menu and keeping their house—mainly the kitchen in the house—as an arena for social practices of the family. These mothers also deal with larger structures of economic and education, which also are resources, though limited for them. They have knowledge on healthy food and geographical proximity should give them access to sources for providing healthy food. Yet for most of them, their limiting structure of education and economy, become barriers for the pre- and post-swallowing of healthy food. The fast pace of urbanization in this rural area, with the following process of compressed modernity (Chang 2010; Ochiai 2014) will pose higher hurdles for them and for the elimination of stunting in the area. Interventions to eliminate stunting should be based on an interdisciplinary approach. In the sociological approach, it is also necessary to further understand the process of (compressed) modernity.

Acknowledgements This study was supported by Direktorat Riset dan Pengabdian Kepada Masyarakat Universitas Indonesia (DRPM UI) under PINTERMIDI UI Research Grant No. 945/SK/R/UI/2019.

References

Abouda, F. E., D. R. Singla, M. I. Nahil, and I. Borisova. 2013. Effectiveness of a parenting program in Bangladesh to address early childhood health, growth and development. *Social Science and Medicine* 97: 250–258. https://doi.org/10.1016/j.socscimed.2013.06.020.

Achadi, E. L., A. Achadi, and T. Aninditha, eds. 2020. *Pencegahan stunting pentingnya peran 1000 hari pertama kehidupan [Prevention of stunting is important in the first 1000 days of life].* London: Rajawali Press.

Chang, K. S. 2010. *South Korea under compressed modernity familial political economy in transition.* London: Routledge.

Delormier, T., K. L. Frohlich, and L. Potvin. 2009. Food and eating as social practice—Understanding eating patterns as social phenomena and implications for public health. *Sociology of Health and Illness* 31 (2): 215–228. https://doi.org/10.1111/j.1467-9566.2008.01128.x.

Faircloth, C., D. M. Hoffman, and L. L. Layne, eds. 2013. *Parenting in global perspective: Negotiating ideologies of kinship, self and politics.* London: Routledge. https://doi.org/10.4324/978 0203103906.

Flynn, J., F. F. Alkaff, W. P. Sukmajaya, and S. Salamah. 2020. Comparison of WHO growth standard and national Indonesian growth reference in determining prevalence and determinants of stunting and underweight in children under five: a cross-sectional study from Musi sub-district. *Research.* https://doi.org/10.12688/f1000research.23156.4.

Giddens, A. 1984. *The constitutions of society.* London: Polity Press.

Giddens, A. 1993. *New rules of sociological method.* London: Polity Press.

Lestari, W., A. Margawati, and Z. Rahfiludin. 2014. Faktor risiko stunting pada anak umur 6–24 bulan di kecamatan Penanggalan kota Subulussalam provinsi Aceh [Risk factors for stunting in children aged 6–24 months in the Penanggalan sub-district, Subulussalam city, Aceh province]. *Jurnal Gizi Indonesia* 3 (1): 37–45. https://doi.org/10.14710/jgi.3.1.126-134.

Maywita, E. 2018. Faktor Risiko Penyebab Terjadinya Stunting Pada Balita Umur 12–59 Bulan Di Kelurahan Kampung Baru Kec. Lubuk Begalung Tahun 2015 [Risk Factors Caused Stunting in Toddlers Aged 12–59 Months in Kampung Baru Kec. Lubuk Begalung in 2015]. *Jurnal Riset Hesti Medan Akper Kesdam I/BB Medan* 3 (1): 56–65. https://doi.org/10.34008/jurhesti.v3i1.24.

Ministry of National Development Planning of the Republic of Indonesia and United Nations Children's Fund. 2017. *Laporan Baseline SDG tentang Anak-Anak di Indonesia* [Baseline Report of SDG about Children in Indonesia]. https://www.unicef.org/indonesia/media/1471/file/SDG% 20Baseline%20report%20Indonesian.pdf

Nugraha, S. Y., N. Fatikhah, S. T. Wahyuni, and N. Saudah. 2019. Influence of social support for families parenting mother in preventing child's stunting. *International Journal of Nursing and Midwifery Science* 3 (3): 122–126. https://doi.org/10.29082/IJNMS/2019/Vol3/Iss3/236.

National Team for the Acceleration of Poverty Reduction. 2017. *100 kabupaten/kota prioritas untuk intervensi anak kerdil* (stunting) [100 priority districts/cities for stunting interventions]. Secretariat of the Vice President of the Republic of Indonesia. http://www.tnp2k.go.id/images/ uploads/downloads/Binder_Volume1.pdf

Ochiai, E. 2014. Leaving the west, rejoining the east? Gender and family in Japna's semi-compressed modernity. *International Sociology* 29 (3): 209–228. https://doi.org/10.1177/026858091453 0415.

Chapter 10
When Free from Stunting Is Not Enough: Cognitive Profile in Children with Stunting and Non-stunting in Nangapanda, East Nusa Tenggara

Sri Redatin Retno Pudjiati, Dhisty Azlia Firnady, and Rizqina Permatasari Ardiwijaya

Abstract Cognitive skills are very important for children's development and performances. Children who have low cognitive profiles tend to show failure both in their study and successful life. One of the factors that significantly impacts children's cognitive profile is nutritional status. East Nusa Tenggara is one of provinces in Indonesia that contribute a high percentage of stunting, which equals 42.6%. This study elaborates the relationship between stunting, parents' stimulation, and cognitive profile in children who live in Nangapanda subdistrict, East Nusa Tenggara. Mix method study design was applied in this research, with purposive sampling. The data were collected from 188 pairs of 5–9-year-old children and their parents. The demographic data were collected by a structured questionnaire by The Southeast Asean Minister of Education Organization (SEAMEO) Recfon. The children's cognitive profile assessed by Cognitive Progressive Matrices (CPM) and the home stimulation assessed by modified Home Observation for Measurement of The Environment: Middle Childhood Questionnaire (HOME-MC). Results show stunting and parents' stimulation were not found correlated with children's cognitive profile. On the other hand, study found that environmental roles such as learning experience, parental education level, and community participation support children's cognitive development. The finding of this result serves as a basis for intervention on families and teachers. Future intervention should be applied to support Nangapanda children's cognitive status.

Keywords Children malnutrition · Cognitive profile · Parents' stimulation · Rural area · Stunting

S. R. R. Pudjiati (✉) · D. A. Firnady · R. P. Ardiwijaya
Department of Psychology, Faculty of Psychology, Universitas Indonesia, Depok, Indonesia
e-mail: retno-pj@ui.ac.id

M. Budianta et al. (eds.), *A Liveable Kampung*, Engaging Indonesia,
https://doi.org/10.1007/978-981-96-7866-2_10

10.1 Introduction

Cognitive abilities are crucial for children, not only to support learning skills, but also to be successful in their life. Cognition will help them to acquire, process, and apply the information from the environment. Having a good cognition supports children to analyze, plan, and create many alternative solutions to solve a problem. Children with different cognitive levels may show different performances in everyday life, learning goal setting, focus and attention, and self-regulation (Phillipson and Phillipson 2012; Goswami 2015).

There are many factors that contribute to children's cognitive level (Neasayan et al. 2019). Social-economic status, family wealth, areas of living, amount of siblings, and parents' education are several factors that impact children's cognitive level (Crookston et al. 2011; Manggala et al. 2018; Titaley et al. 2019). Unfortunately, biological factors such as children's diet and nutrition tend to be neglected as a primary factor that contribute to children's cognitive function (Jirout et al. 2019). Children with poor early nutrition intake are at risk of failing to reach expected linear growth milestones, leading to stunting, long-term cognitive and motor deficits, and eventually worse economic outcomes in the future (Dewey and Begum 2011; Galasso et al. 2017). In some settings, concurrent nutritional status may be at least as important as early nutritional status in predicting children's cognitive performance (Jirout et al. 2019).

Nutritional problems remain alarming. Three indicators of malnutrition are stunting, underweight, and wasting. In 2018, 151 million children under 5 years old are stunting (UNICEF, WHO, The World Bank 2018). Stunting is a status that is given to the child whose height is shorter than another equal-aged-child (WHO 2010). Many studies show that stunting hinders individual development and performances. Children who are labeled as stunting show significant declines in learning ability, academic score, and chronic disease. Pemerintah Provinsi East Nusa Tenggara (East Nusa Tenggara), Dewan Pertahanan Pangan, Kementerian Pertanian, and World Food Program (2015) already stated that stunting is a significant obstacle to economic growth and national development. In order to that, stunting is considered to be a global public health problem among children and being a big challenge to many developmental countries (WHO 2010; Dewey and Begum 2011; de Onis and Branca 2016; Kementerian Desa et al. 2017; Titaley et al. 2019).

East Nusa Tenggara is one of the provinces with the highest percentages of stunting in Indonesia, which is around 42.6% (Badan Pusat Statistik 2019a; b). One of the provinces in East Nusa Tenggara is Ende District. In 2013, the prevalence of stunting in Ende is 35.99% (Pemerintah Provinsi East Nusa Tenggara (East Nusa Tenggara), Dewan Pertahanan Pangan, Kementerian Pertanian, and World Food Program 2015). According to WHO (2010) classification, those conditions classified as a severe health problem (stunting > 20%),

As stated in UNICEF conceptual framework of the determinant of children undernutrition, many studies show that most stunting cases around the world happened in low-income and non-wealthy families (Willey et al. 2009; Dewey and Begum 2011;

UNICEF 2015; Pacheco et al. 2017). Nangapanda, one of the subdistricts in Ende, is a coastal-rural-area with a population of approximately 22,000 people divided over 29 villages (BPS Kabupaten Ende 2019a, b). Most of the citizens were farmers, fishers, and civil servants (BPS Kabupaten Ende 2019a, b). Based on previous studies, Nangapanda has a high stunting prevalence, that is 35.9% (Pemerintah NTT, Dewan Ketahanan Pangan, WFP 2013), which this percentage is considered as a severe problem based on WHO. According to the data, the study is considered to be conducted in Nangapanda subdistrict, Ende, East Nusa Tenggara.

Nangapanda is not only surrounded by the sea but also mountains. Unfortunately, Nangapanda is classified as a dry area. In some areas, it is still difficult to get hygienic water. Plant variations that can be cultivated in Nangapanda are limited, and it also impacts everyday dietary menus that do not vary. Health-care facilities are also very limited; there is no public hospital around Nangapanda subdistrict. Education facilities are limited to primary schools (30 schools). There are only 8 junior high schools and 2 senior high schools around Nangapanda (BPS Kabupaten Ende 2019a, b). Students of junior or senior high school must go out of town to have an education because. Citizens who live there are middle (senior high school) to low education (elementary school), or not finishing/going to school.

Nowadays, several programs have been implemented to reduce stunting, such as medication intake, economic development, and agriculture intervention (Pemerintah Provinsi East Nusa Tenggara, Dewan Pertahanan Pangan, Kementerian Pertanian, and World Food Program 2015; Rokx et al. 2018). On the other hand, data still shows a great prevalence of stunting in NTT. Home environments such as childcare and stimulation are also found as factors that are related to stunting. Research shows that parents who come from low-income families could not give optimum stimulation and use the resources to parenting their children (Beal et al. 2018).

The poverty experienced by children will have a huge negative impact on their future development (Li et al. 2017). Not only in the feeding setting, from a preliminary study that was already conducted in Nangapanda, it shows that 17.3% parents tend to use harsh discipline toward their children and 25.06% of parents only support children's basic needs. This situation certainly supports low cognitive profile, behavior and emotional problems, and developmental delay in certain areas (Augusti and Melinder 2013; Maguire et al. 2015).

According to the previous explanation, both stunting and non-stunting children get the same treatment from their parents. This research was conducted to evaluate correlation between parents' stimulation and children's cognitive profile, both in stunting and non-stunting children. The aim of this research is to provide information as a basis for making an intervention for school age children living in poverty and rural areas.

10.2 Methods

10.2.1 Research Design

Mix method study design was applied in this research. This study analyzed both quantitative and qualitative data. The correlation of parents' stimulation toward cognitive profile in children with and without stunting were compared. Multiple regression was applied to evaluate the contribution of stunting, parents' stimulation, and demographic data toward children's cognitive profile.

10.2.2 Participants and Data Collection

In this research, a purposive sampling method was applied. The data were collected from 188 pairs of 5–9 years old children and their parents who reside in Nangapanda Subdistrict, Ende, East Nusa Tenggara, Indonesia. The data were collected from September 30, 2019, until October 14, 2019.

10.2.3 Measurement

10.2.3.1 Demographic Data

Demographic data of each household member were collected using a structured questionnaire developed by The Southeast Asian Ministers of Education Organization (SEAMEO). Children's gender and age, areas of living (rural or urban area), maternal and paternal age, maternal and paternal education level, and maternal and paternal occupation were assessed.

10.2.3.2 Children's Cognitive Profile

The children's cognitive profile as the main outcome was assessed by Colored Progressive Matrices (CPM) and academic test. CPM analyzes one of intelligence aspects, for instance non-verbal reasoning for children aged 5–11 years old. Split-half reliability score of CPM ranging from 0.65 to 0.94, and the test–retest reliability score was between 0.71 and 0.93. Validity coefficient of CPM compared to intelligence test ranging from 0.50 to 0.80 (Raven et al. 1986; Sattler 1992). Academic test was executed to evaluate children's comprehension of basic subjects in school, that is Bahasa Indonesia and Mathematics. This test was conducted based on the 2013 Indonesian National Curriculum.

10.2.3.3 Parents Stimulation

Parents' daily stimulation was evaluated by the modification of Home Observation for Measurement of The Environment: Middle Childhood Questionnaire (HOME-MC). Modified HOME-MC consists of six subscales, that is: (1) Emotional and verbal responsibility; (2) Encouragement of maturity; (3) Emotional climate; (4) Learning material and opportunities (e.g. the number of children's books in the household); (5) Enrichment; and (6) Family companionship (e.g. parents' engagement in play activities with the children). Internal consistency and inter-observer agreement reliability coefficients of HOME-MC ranging from 0.52 to 0.80 for the subscales (Bradley et al. 1988; Caldwell and Bradley 2016).

10.2.3.4 Nutritional Status

Nutritional status was assessed in children anthropometry assessments. Stunting will be determined if children's height-for-age-z-score (HAZ) < – 2.00 (Gibson 2005).

10.2.4 Method Analysis

The data were double cleaned and checked for missing values and outliers. Descriptive analyses were performed using frequencies and percentages for categorical variables. Associations between participant characteristics, stunting, cognitive profile, and parents' stimulation were determined by Pearson's product-moment correlation coefficient and multiple regression (Tables 10.1, 10.2, 10.3).

Table 10.1 Data descriptive

	N	Range	Min	Max	Sum	Mean	Std. deviation
Age (month)	188	42	71	113	16,931	90.06	5.641
Gender	188	1	1	2	269	1.43	0.497
Father's occupation	188	76	1	77	730	3.88	6.746
Mother's occupation	188	76	1	77	705	7.98	11.274
Father's education	188	5	1	6	680	3.62	1.233
Mother's education	188	5	1	6	1500	3.75	1.231
CPM score	188	32	0	32	2837	15.09	5.487
Academic test score	188	18	1	19	1855	9.87	4.585
HOME score	188	31	13	44	5081	27.03	5.883
Stunting status	188	1	0	1	120	0.64	0.482

Table 10.2 Inter-variable correlation

	Age (month)	Gender	Father's occupation	Mother's occupation	Father's education	Mother's education	CPM score	Academic test score	HOME score	Stunting status
Age (month)	1									
Gender	−0.066	1								
Father's occupation	−0.078	−0.062	1							
Mother's occupation	0.134	0.076	0.047	1						
Father's education	0.019	0.140	0.094	−0.018	1					
Mother's education	−0.006	0.195**	0.167*	0.023	0.691**	1				
CPM score	0.174*	0.129	0.087	0.120	0.243**	0.277**	1			
Academic test score	0.253**	0.263**	−0.045	0.026	0.279**	0.226**	0.403**	1		
HOME score	0.085	0.047	−0.058	0.059	0.166*	0.231**	0.074	0.066	1	
Stunting status	−0.063	0.093	0.099	−0.049	0.092	0.093	0.130	0.031	−0.096	1

Table 10.3 Regression analysis

Variable	*B*	*R*	*R2*	Δ*R*	*t*	Variable	*B*	*R*	*R2*	Δ*R*	*t*
Level 1		**0.470****	**0.221****	**0.186****		**Level 2**		**0.471****	**0.221****	**0.183****	
Age (month)	0.081				0.234	Age (month)	0.085				0.212
Gender	0.082				0.916	Gender	− 0.021				0.979
Father's occupation	0.060				0.279	Father's occupation	0.053				0.342
Mother's occupation	0.045				0.172	Mother's occupation	0.047				0.147
Father's education	0.109				0.792	Father's education	0.083				0.840
Mother's education	0.753				0.076	Mother's education	0.729				0.085
Academic test score	0.402				0.000**	Academic test score	0.402				0.000**
HOME score	0.000				0.997	HOME score	0.011				0.860
						Stunting status	1.334				0.110

* Significant at $p < 0.05$
** Significant at $p < 0.01$
Level 1 indicates that 18.6% of variation in cognitive variables is explained by covariate variables and HOME score (R = .221, adjusted R = 0.186, p.01)
Level 2 shows that by controlling covariate variables and stunting, there is 18.3% increase in variance of cognitive performance (R2 = 0.221, adjusted R = 0.183, p.01)

10.3 Results

Research found that 64.4% of children have typical nutritional status and 35.6% of children were stunting. Cognitive profile was correlated with academic test ($r = 0.403$, $p < 0.01$). Academic tests correlated with children's age ($r = 0.250$, $p < 0.01$). Father education level found significantly correlates with cognitive profile ($r = 0.243$, $p < 0.01$), academic test ($r = 0.279$, $p < 0.01$), and HOME stimulation ($r = 0.166$, $p < 0.05$). Mother education level also correlated with cognitive profile ($r = 0.277$, $p < 0.01$), academic test ($r = 0.226$, $p < 0.05$), and HOME stimulation ($r = 0.231$, $p < 0.01$).

The result did not find any correlation and contribution between stunting status and cognitive profile ($r = 0.130$, $p > 0.01$, 0.05). The result also did not find any correlations and contribution between cognitive profile and HOME stimulation ($r = 0.074$, $p > 0.01$, 0.05). In the other words, there might be other variables that could affect Nangapanda children's cognitive profile that were not studied yet in this research.

10.4 Discussion and Conclusion

Cognitive profile and academic test were found to be correlated with children's age. It shows that children's learning ability is affected by environmental factors, such as learning experience. The more children get the opportunity to learn, the better their academic performance. Based on brain imaging research, it suggests that children who grow up in a disadvantaged environment, such as low socioeconomic status, cause the brain to develop differently especially in brain areas related to learning and memory, language ability, and socioemotional (Gianaros 2011; Hanson et al. 2011; Noble et al. 2012).

Based on the result, nutritional status was not found to be correlated with the children's cognitive profile. This result contrasts with the several studies that suggest height-for-age-*z*-score (HAZ) had significant positive associations with learning performances among children (Acham et al. 2008; Abebe et al. 2017). Interesting finding was found in this research, that is both stunting and typical children living in Nangapanda have a low ability in analyzing non-verbal stimuli, proven by 71.3% of all participants who got 'below average' scores on the CPM test. In the other words, stunting is not only a single factor that contributes to children's cognitive profile. According to Miller et al. (2019), children's development is strongly influenced by their life context. Study from Hermida et al. (2019) shows children from rural areas significantly have a lower score of non-verbal intelligence test. Tine (2014) also found a unique finding that low-income rural children had worse visuospatial working memory score than verbal working memory score. It is undeniable that children who live in rural areas face several disadvantages, and it creates a big gap in their cognitive aspects compared to other advantageous children (Miller et al. 2019).

The level of parents' education (both father and mother) has a significant correlation with cognitive profile, academic test, and parents' stimulation. In this study, 68.6% of parents have a low-education level. Parents who have lower education could not engage in stimulating, teaching, and supporting children. As they grow, low-education parents tend to show less useful information to help their children's early skill. In line with that, children of low education and poor parents tend to have less family investment in education (deCoulon et al. 2011; Lucassen et al. 2015; Miller et al. 2019; Berkes et al. 2019).

In contrast, a significant correlation between parents' stimulation and children's cognitive profile both in children with or without stunting were not found. It shows that daily stimulation is also not enough to support children's cognitive profile. There are some other factors that may contribute to this situation, namely the low quality of parent–child relationship. Parents could not use their resources (i.e., knowledge, traditional things, toys, etc.) to provide optimum stimulation. For example, to give proper rules and challenges matched with the child's needs; to build an effective communication; and to show a warm affection to the child.

Based on qualitative data, there is a tendency that almost all parents in Nangapanda use verbal and physical violence toward their children. Unfortunately, the parents' stimulation instrument that was used in this research does not measure harmful parenting specifically. Previous research shows that parental warmth and lovingness correlated with children's performance. Sensitive and warm parents are able to understand the children's needs and provide enough stimulation for the children (Towe-Goodman et al. 2014; Berkes et al. 2019). In other ways, all forms of negative discipline have a significant impact on children's development across the lifespan, including cognitive function (Gershoff and Grogan-Kaylor 2016). Based on that condition, despite the nutrition status being typical, children who get a wrong treatment from their significant others will potentially show worse performance compared to stunting children. In line with the result, it was found that 62.7% of 118 children have low average cognitive performance. Further research needed to study the impact of harmful parenting behavior toward the children's cognitive development.

The ideal stimulation of parents to support children's cognitive skill can be provided through serving interaction, playing activity, and facilitating developmental needs such as reading books together and teaching the children (Shonkoff and Phillips 2000; Bradley et al. 2011). On the other hand, low-income families find difficulties in providing both investment and proper stimulation to their children (Oxford and Lee 2011). From the HOME-MC questionnaire, it was found that low-income parents in Nangapanda have limited resources, so they cannot provide some stimulating things such as story books, educational toys, radio, television, and other electronic devices to support their children's general knowledge.

In this study, 59.8% of participants show a high level of children's participation in their community that can be found in the last domain of HOME-MC. This is one of the protective factors that can support children's cognitive skills and other developmental aspects. Their involvement in various activities in the environment becomes the stimulation to build their competence. Children can play freely with their friends, make their own toys from the materials found in the environment, and socialize with other people. It can increase creativity and the ability to work practically on the child. Some children have small groups to do their homework. Qualitatively, we also found that older children teach the younger children. The study groups can facilitate the transfer of knowledge between children. Research concludes that higher levels of smartphone use can predict children's poor academic performance (Beland and Murphy 2014). It means the lack of gadget use can help the children undergo the optimum learning process from the environment.

Uniquely, based on the researchers' observation, teacher's involvement also plays an important role as a resource toward children's academic performance. Most of the Nagapanda's parents feel they do not have sufficient knowledge and power, so they demand the teachers to fully educate and support their children, not only in cognitive aspects but also in other aspects such as social skill and emotional development. In line with this finding, research from Sawai et al. (2017) also shows that socioeconomic status negatively correlated with the mother's self-esteem.

Acknowledgements This study was conducted under Penelitian Internasional Terpadu Multi Interdisiplin Universitas Indonesia (PINTER MIDI UI). The researchers sincerely thank Laras Sekarasih, M.S., Ph.D. for the guidance and support.

References

Abebe, F., A. Geleto, L. Sena, et al. 2017. Predictors of academic performance with due focus on undernutrition among students attending primary schools of Hawa Gelan district, Southwest Ethiopia: A school based cross sectional study. *BMC Nutrition* 3: 30. https://doi.org/10.1186/s40795-017-0138-2.

Acham, H., J. K. Kikafunda, S. Oluka, M. K. Malde, and T. Tylleskar. 2008. Height, weight, body mass index and learning achievement in Kumi district, East of Uganda. *Scientific Research and Essay* 3 (1): 1–8. https://doi.org/10.5897/SRE.9000544.

Augusti, E. M., and A. Melinder. 2013. Maltreatment is associated with specific impairments in executive functions: A pilot study. *Journal of Traumatic Stress* 26 (6): 780–783. https://doi.org/10.1002/jts.21860.

Badan Pusat Statistik Kabupaten Ende. 2019. *Kabupaten Ende dalam angka 2019* (*Ende regency in figures 2019*). CV Unit Print Ende.

Badan Pusat Statistik Kabupaten Ende. 2019. *Kecamatan Nangapanda dalam angka 2019* (*Nangapanda district in figures 2019*). CV Unit Print Ende.

Beal, T., Tumilowicz, A., Sutrisna, A., Izwardy, D., & Neufeld, L. M. 2018. A review of child stunting determinants in Indonesia. *Maternal & Child Nutrition* 14 (4): e12617. https://doi.org/10.1111/mcn.12617.

Beland, L. P. & Murphy, R. 2016. Ill Communication: Technology, distraction & student performance. *Labour Economics* 41: 61–76.

Berkes, J., A. Raikes, A. Bouguen, and D. Filmer. 2019. Joint roles of parenting and nutritional status for child development: Evidence from rural Cambodia. *Developmental Science* 22 (5): 1–19. https://doi.org/10.1111/desc.12874.

Bradley, R. H., B. M. Caldwell, S. L. Rock, H. M. Hamrick, and P. Harris. 1988. Home observation for measurement of the environment: Development of a home inventory for use with families having children 6 to 10 years old. *Contemporary Educational Psychology* 13 (1): 58–71. https://doi.org/10.1016/0361-476X(88)90006-9.

Bradley, R. H., L. McKelvey, and L. Whiteside-Mansell. 2011. Does the quality of stimulation and support in the home environment moderate the effect of early education programs? *Child Development* 82 (6): 2110–2112. https://doi.org/10.1111/j.1467-8624.2011.01659.x.

Caldwell, B. M., & Bradley, R. H. (2016). *Home observation for measurement of the environment: Administration manual*. Family & Human Dynamics Research Institute, Arizona State University.

Crookston, B. T., K. A. Dearden, S. C. Alder, C. A. Porucznik, J. B. Stanford, R. M. Merrill, T. T. Dickerson, and M. E. Penny. 2011. Impact of early and concurrent stunting on cognition. *Maternal and Child Nutrition* 7 (4): 397–409. https://doi.org/10.1111/j.1740-8709.2010.00255.x.

de Coulon, A., E. Meschi, and A. Vignoles. 2011. Parents' skills and children's cognitive and non-cognitive outcomes. *Education Economics* 19 (5): 451–474. https://doi.org/10.1080/09645292.2010.511829.

de Onis, M., and F. Branca. 2016. Childhood stunting: A global perspective. *Maternal and Child Nutrition* 12 (1): 12–26. https://doi.org/10.1111/mcn.12231.

Dewey, K. G., and K. Begum. 2011. Long-term consequences of stunting in early life. *Maternal and Child Nutrition* 7 (3): 5–18. https://doi.org/10.1111/j.1740-8709.2011.00349.x.

Galasso, E., A. Wagstaff, S. Nadeau, and M. Shekar. 2017. *The economic costs of stunting and how to reduce them*. World Bank Policy Research Note. PRN/17/05. http://pubdocs.worldbank.org/en/536661487971403516/PRN05-March2017-Economic-Costs-of-Stunting.pdf

Gershoff, E. T., and A. Grogan-Kaylor. 2016. Spanking and child outcomes: Old controversies and new meta-analyses. *Journal of Family Psychology* 30 (4): 453–469. https://doi.org/10.1037/fam0000191.

Gianaros, P. J. 2011. Socioeconomic health disparities: A health neuroscience and lifecourse perspective. *Psychological Science Agenda* 25 (1): 345–356.

Gibson, R. S. 2005. *Principles of nutritional assessment*, 2nd ed. Oxford: Oxford University Press.

Goswami, U. 2015. *Children's cognitive development and learning*. Cambridge Primary Review Trust. https://www.talk4writing.com/wp-content/uploads/2019/04/COMPLETE-REPORT-Goswami-Childrens-Cognitive-Development-and-Learning.pdf

Hanson, J. L., A. Chandra, B. L. Wolfe, et al. 2011. Association between income and the hippocampus. *PLoS ONE* 6 (5): e18712. https://doi.org/10.1371/journal.pone.0018712.

Hermida, M. J., et al. 2019. Risks for child cognitive development in rural context. *Frontiers in Psychology* 9: Article 2735. https://doi.org/10.3389/fpsyg.2018.02735.

Jirout, J., et al. 2019. How lifestyle factors affect cognitive and executive function and the ability to learn in children. *Nutrients* 11 (8): 1953. https://doi.org/10.3390/nu11081953.

Kementerian Desa, Pembangunan Daerah Tertinggal, dan Transmigrasi. 2017. *Buku Saku Desa dalam Penanganan Stunting* [Village Pocket Book in Handling Stunting]. https://siha.depkes.go.id/portal/files_upload/Buku_Saku_Stunting_Desa.pdf

Li, C., X. Yin, and S. Jiang. 2017. Effects of multidimensional child poverty on children's mental health in Mainland China. *Journal of Health Psychology* 25 (3): 400–415. https://doi.org/10.1177/1359105317718379.

Lucassen, N., et al. 2015. Executive function in early childhood: The role of maternal and paternal parenting practices. *British Journal of Developmental Psychology* 33 (4): 489–505. https://doi.org/10.1111/bjdp.12112.

Maguire, S. A., et al. 2015. A systematic review of the emotional, behavioural and cognitive features exhibited by school-aged children experiencing neglect or emotional abuse. *Child: Care Health and Development* 41 (5): 641–653. https://doi.org/10.1111/cch.12227.

Manggala, A. K., K. W. Kenwa, M. M. Kenwa, A. A. Sakti, and A. A. Sawitri. 2018. Risk factor of stunting in children age 24–59 months. *Paediatrica Indonesiana* 58 (5): 205–212. https://doi.org/10.14238/pi58.5.2018.205-12.

Miller, P., E. Votruba-Drzal, and R. Coley. 2019. Poverty and academic achievement across the urban to rural landscape: Associations with community resources and stressors. *RSF: the Russell Sage*

Foundation Journal of the Social Sciences 5 (2): 106–122. https://doi.org/10.7758/RSF.2019.5.2.06.
Nesayan, A., M. Amani, and R. A. Gandomani. 2019. Cognitive profile of children and its relationship with academic performance. *BCN 2019* 10 (2): 165–174. https://doi.org/10.32598/bcn.9.10.230.
Noble, K. G., S. M. Houston, E. Kan, et al. 2012. Neural correlates of socioeconomic status in the developing human brain. *Developmental Science* 15 (4): 516–527. https://doi.org/10.1111/j.1467-7687.2012.01147.x.
Oxford, M. L., and J. O. Lee. 2011. The effect of family process on school achievement as moderated by socioeconomic context. *Journal of School Psychology* 49 (5): 579–612. https://doi.org/10.1016/j.jsp.2011.06.001.
Pacheco, C. D. R., Picauly, I., & Sinaga, M. 2017. Health, food consumption, social economy, and stunting incidence in Timor Leste. *KEMAS: Journal Kesehatan Masyarakat* 12 (2): 168–177. https://doi.org/10.15294/kemas.v12i2.11248.
Pemerintah Provinsi East Nusa Tenggara, Dewan Pertahanan Pangan, Kementerian Pertanian, and World Food Programme. 2015. *Peta Ketahanan dan Kerentanan Pangan East Nusa Tenggara* [East Nusa Tenggara Food Security and Vulnerability Map]. https://documents.wfp.org/stellent/groups/public/documents/ena/wfp285131.pdf?iframe
Phillipson, S., and S. N. Phillipson. 2012. Children's cognitive ability and their academic achievement: The meditation effect of parental expectation. *Asia Pacific Education Review* 13: 495–508. https://doi.org/10.1007/s12564-011-9198-1.
Raven, J. C., J. H. Court, and J. Raven. 1986. *Manual for raven's progressive matrices and vocabulary scales*. London: Lewis.
Rokx, C., A. Subandoro, and P. Gallagher. 2018. *Aiming high Indonesia's ambition to reduce stunting*. London: The World Bank.
Sarma, M. S. G., D. G. N. G. Wijesinghe, and T. Sivananthawerl. 2013. The effects of nutritional status on educational performance of primary school children in the plantation sector in Nuwara Eliya educational zone. *Tropical Agricultural Research* 24 (3): 203–214. https://doi.org/10.4038/tar.v24i3.8005.
Sattler, J. M. 1992. *Assessment of children*. London: Jerome M. Sattler Publisher Inc.
Sawai, J. P., F. Bahari, H. Ibrahim, and Z. K. Rejab. 2017. The relationship between economic hardship, self-esteem, and parental behavior among low-income mothers. *Jurnal Kemanusiaan, 9*(2), pp. 59–76. https://jurnalkemanusiaan.utm.my/index.php/kemanusiaan/article/view/104
Septika, I. M. W. D. 2019 *Opini: Melawan Stunting itu Penting* [Opinion: Fighting Stunting is Important]. Pos-Kupang.com https://kupang.tribunnews.com/2019/02/13/melawan-stunting-itu-penting
Shonkoff, J. P., and D. A. Phillips. 2000. *From neurons to neighborhoods: The science of early childhood development*. London: National Academy Press.
Titaley, C. R., I. Ariawan, D. Hapsari, A. Muasyaroh, and M. J. Dibley. 2019. Determinants of the stunting of children under two years old in Indonesia: A multilevel analysis of the 2013 Indonesia basic health survey. *Nutrients* 11 (1106): 1–13. https://doi.org/10.3390/nu11051106.
Tine, M. 2014. Working memory differences between children living in rural and urban poverty. *Journal of Cognition Development* 15 (4): 599–613. https://doi.org/10.1080/15248372.2013.797906.
Towe-goodman, N. R., M. Willoughby, C. Blair, H. C. Gustafsson, W. R. Mills-koonce, and M. J. Cox. 2014. Fathers' sensitive parenting and the development of early executive functioning. *Journal of Family Psychology* 28 (6): 867–876. https://doi.org/10.1037/a0038128.
UNICEF. 2015. *Unicef's approach to scaling up nutrition for mothers and their children*. https://www.unicef.org/nutrition/files/Unicef_Nutrition_Strategy.pdf.
UNICEF, WHO, The World Bank. 2018. *Joint child malnutrition estimates 2018 edition*. https://www.who.int/nutgrowthdb/2018-jme-brochure.pdf
WHO. 2010. *Nutrition landscape information system (NLIS) country profile indicators: interpretation guide*. https://apps.who.int/iris/handle/10665/44397

Willey, B. A., Cameron, N., Norris, S. A., et al. 2009. Socio-economic predictors of stunting in preschool children: A population-based study from Johannesburg and Soweto. *South African Medical Journal* 99 (6): 450–456.

Chapter 11
Desegregation in Action: Understanding Livability from Kampung Advocacy by Chinese-Indonesian Urban Activists

Rita Padawangi

Abstract In this chapter, I re-examine the concept of livability through the experiences of Chinese-Indonesian activists in urban kampung advocacy. The purpose is to discuss the extent to which collective governance and collective action in kampung can stand as an inspiration of a society who can work together for common good. Studies have pointed to the contradictory relationship between urban development and kampung *kota* in Indonesia, as old settlements may be forced to make way for new buildings, infrastructures, and population. With the exception of studies of old Chinese quarters, discourses on Chinese-Indonesians are rarely situated within urban kampungs, and stereotypes of Chinese-Indonesians as entrepreneurs perpetuate the distance between Chinese-Indonesianness and kampung. The objective of this chapter is threefold: (1) to enrich the discussion on livability from the lens of kampung, particularly by challenging the popular assumption of livability as services and infrastructures and (2) addressing the ethnic segregation and fragmentation, which also take place in kampung as a result of urban expansion. Finally, the third objective is to demonstrate the lens of kampung in the issue of discrimination and segregation. How do Chinese-Indonesian activists perceive their relationship with urban kampung? What are the grounds of their involvement in urban kampung activism? How do experiences in urban kampung activisms affect their understanding of their Chinese-Indonesianness vis-à-vis kampung as a social, cultural, and spatial unit? I rely on in-depth interviews of Chinese-Indonesian kampung activists in Jakarta, Bandung, and Surabaya. Urban kampung advocacies are spaces in which Chinese-Indonesian activists form relationships and develop spatial rootedness in the neighborhoods. As their primary residences are rarely in urban kampung, the potential of desegregation remains limited, but interactions through urban kampung activism are in themselves processes in pursuit of livability, and for the activists a re-negotiation of the sense of Chinese-Indonesianness.

Keywords Activists · Advocacy · Segregation · Liveability

R. Padawangi (✉)
College of Interdisciplinary & Experiential Learning, Singapore University of Social Sciences, Singapore, Singapore
e-mail: ritapadawangi@suss.edu.sg

M. Budianta et al. (eds.), *A Liveable Kampung*, Engaging Indonesia,
https://doi.org/10.1007/978-981-96-7866-2_11

11.1 Introduction

Given the contradictory relationship between urban development and kampung in urbanizing Indonesia, why does kampung remain important, and why is it central to achieving a livable city? As the previous chapters have shown, even when kampung face challenges due to lack of service provision, infrastructures as well as support from the state in provision of public goods, kampung are places where we can observe collective governance and collective action in everyday life, including to address these urban development challenges. But to answer the extent to which kampung remains important in understanding livability of the city, it is necessary to look beyond kampung to understand how kampung is important to the city at-large and not just for kampung itself. Hence, my discussion in this chapter focuses on narratives of those who do not live in kampung—the outsiders—but give their efforts to advocate for kampung in urban policies, particularly against forced evictions.

Therefore, I choose to study a specific group among urban activists: the Chinese-Indonesians who have become advocates against kampung evictions. Among all aspects about kampung, why do I choose to look at Chinese-Indonesians? I do not usually study Chinese-Indonesians, as I thought there are already many discussions on the history, social issues, and identity politics in relation to Chinese-Indonesianness. But there is one thing that drove me to conduct this small study about kampung advocacy: I believe this aspect has been lacking, both in understanding kampung and in understanding the social role of Chinese-Indonesians in the city. I believe this is a very important issue to be addressed in light of escalations of ethnic tensions during the heated Jakarta gubernatorial elections in 2017. Examining narratives of Chinese-Indonesian kampung activists will help to understand how collective governance and collective action in kampung generate solidarity among its residents.

Many Indonesian studies have highlighted discrimination toward the Chinese, from formal bureaucratic procedures to gestures of different treatment in everyday life. However, the studies do not address the spatial aspects of discrimination. With the exception of studies of old Chinese quarters, discourses on Chinese-Indonesian communities are rarely situated within urban kampung because they are generally viewed as highly affluent people who enjoy secured and luxurious life within gated housing complexes. Such a stereotype perpetuates the perceived distance between Chinese-Indonesians and kampung. As a result, although Chinese-Indonesians are still present among urban kampung activists, Chinese-Indonesianness is rarely associated with advocacies alongside urban kampung struggles.

How do Chinese-Indonesian activists perceive their relationship with urban kampung? What are the grounds of their involvement in urban kampung activism? How do experiences in urban kampung activism affect their own understanding of Chinese-Indonesianness? Focusing on Chinese-Indonesian kampung activists will also directly critique the lack of spatial perspective in discussing the escalation of ethnic tensions. The objective of this chapter is threefold: (1) to enrich the discussion on livability from the lens of kampung, particularly by challenging the popular assumption of livability as services and infrastructures and (2) addressing the ethnic segregation and fragmentation, which also take place in kampung as a result of urban expansion. Urban expansion in this regard does not only pose challenges that relate to service sectors such as sanitation, health, education. In fact, the political economy of urban expansion in Indonesia has subjected kampung to becoming sites of these fragmentations and segregation. Finally, the third objective is to demonstrate the lens of kampung in the issue of discrimination and segregation. Learning closely from kampung allows us to move away from perpetuating the image of Chinese-Indonesians as victims of discrimination and segregation. Kampung allows us to pay attention to the human agency of Chinese-Indonesians in addressing the situation beyond the boundaries of ethnic communities. Thus, kampung as a mixed district that includes Chinese-Indonesians is an important spatial unit in contextualizing that agency.

11.2 Kampung

What is kampung? Indonesians seem to use the term lightly on so many occasions, but what does it refer to? The term "kampung" in Indonesian conversations, in the context of a place of origin, refers to something rather nostalgic (*kampung halaman*, for example, means hometown). In the context of urbanization, the term "kampungan" has a more derogatory connotation, referring to behavior associated with being from a kampung that is unsuitable to modernity and the required habits of urban living. Literally, the word "kampungan" means behaving like people from kampung, who are unfit for the fast-paced, efficient, and industrial city.

Such a negative connotation, however, contradicts the historical background of cities in Southeast Asia that mostly started with kampung as settlements. In his studies of various cities in Southeast Asia, Imran bin Tajudeen (2019) suggested that the term "campong," which is pronounced similarly to "kampung," has been used by the Portuguese, Dutch, and English in the colonial period to refer to urban wards. The term represents neighborhood as understood by European colonial authorities in the region. Most cities in Southeast Asia, if not all, originated from these settlements and eventually developed with different scales according to the layers of social and economic transformations in their histories.

Kampung, therefore, is historical, but it is also contemporary. In today's context, a kampung *kota* or urban kampung is a district, ward, neighborhood in cities of Southeast Asia, in which there is a sense of residents coming together to form a community and celebrate major festivities together. This sense of neighborliness is a contemporary manifestation of that contradiction with an industrialized society, which is associated with nuclear family settings in their private homes as spaces of reproduction that do not require much interaction with their neighbors. But as a settlement, most governments and urban planners in the modern era—postcolonial industrialization—paint kampung as an object that needs to be revitalized, improved, or even demolished for the sake of development (Kuntjoro-Jakti 1986; Silas 1996). Such a condescending view toward kampung is not only happening in cities in Indonesia, but peppered throughout Southeast Asia. In Singapore, a country with the highest GDP in the region, the city demolished all but one kampung to make way for high-rise housing, industrial areas, parks, and many kinds of space-intensive infrastructures. The image of the kampung is rarely associated with modernity.

11.3 Densification of Cities

As Southeast Asia's economies intensively industrialized in the 1970s, populations of cities and their urban regions grew exponentially. Jakarta's population nearly doubled from 2.6 million in 1960 to 4.8 million in 1975. However, housing provision by the government had been few and far in between, resulting in densification of kampungs as places that allow for organic growth of population and relatively versatile building modifications. Such densification would have an impact on the capacity of the public infrastructures and environmental qualities that were outpaced by the population growth. In comparison, a city like Singapore also saw a significant population growth in the same period of time, but only doubled the population in more than 20 years' time. From 1.6 million in 1960, Singapore's population reached approximately 2.6 million in 1985. The pattern of population increase through migration is different, though, since migrants who entered Singapore would have to pass through the national border, placing an additional bureaucratic requirement compared to in-country migration.

Removal of kampung to make way for new development projects in the city becomes a typical occurrence, as Indonesia's cities densify. In Singapore the densification of kampung did not take place as the population increase was paralleled by the demolition of kampung and relocation of the inhabitants to multi-story public housing compounds. In all cities of Indonesia, public housing projects do not dominate housing supply and, therefore, the increase of urban population resulted in the densification of kampung (Kusno 2014). Large-scale housing projects that took place in the 1980s during the real estate boom were carried out by private developers. While these developments catered to some of the housing needs of the middle class, kampung remains an affordable housing option for migrants and new families, especially with their strategic locations in the city and hence the proximity to established urban services and jobs, compared to private housing projects in the outskirts that would have required them to commute.

11.4 Ethnic Transformation of Kampung

The booming real estate development by private developers in the 1980s in the outskirts of cities—primarily cities in Java—and in satellite towns occurred at the peak of Soeharto's presidency. New towns in the Jakarta Metropolitan Region, which includes its surrounding cities: Bogor, Depok, Tangerang, and Bekasi, started to develop in the early 1980s, and by mid-1990s there were almost 25 large projects, ranging from 500 to 6000 hectares (Firman 2004, p. 350). Acknowledging the impact of such projects that perpetuate segregation, Firman (2004) called this phenomenon "self-segregation" or "voluntary segregation," because upper middle- and high-income buyers of houses in these gated communities decided to reside there by choice.

However, in the case of Chinese-Indonesians, their choice is also affected by the persistent negative stereotyping of their ethnic group, combined with their own negative stereotyping of non-Chinese. Therefore, their choices of residence were constrained by concerns of safety, and for some, superiority. Although ethnic and racial diversity had been recognized by Indonesia's national(ist) leaders as one of the challenges faced by the nation, it was under Soeharto's reign that the solidification of a systematic discrimination toward the Chinese ethnic group began to be carried out. Post-1965 massacre, the Chinese ethnic group had been subjected to various additional obligations related to their Indonesian citizenship status, such as the necessity to have a citizenship certificate (*Surat Bukti Kewarganegaraan Indonesia*) and other requirements of paperwork for various formal administrative procedures.

Historians have documented the spatial segregation of ethnic groups in the management of cities during the colonial era, i.e., the *Wijkenstelsel* zoning system and the *Passenstelsel*, which specifically targeted the Chinese in terms of gaining a permit to legally allow them to go out of their quarters. While these result in traces of ethnic segregation in contemporary cities, both in terms of buildings and among residents of different ethnic quarters, kampung at the time became a mixed zone, as local leaders were given the autonomy to oversee who could reside there, as stated by Perkasa et al. (2022, p. 367).

That was in the colonial era. Fast forward to 2019, the picture was significantly different. There are still Chinese-Indonesian families in kampungs, but the atmosphere has changed. In Kampung Peneleh, Surabaya, Farida (2019), in her ethnography, documented how an elderly Chinese-Indonesian woman recounted her experience when she first moved into the kampung. The reason why she did so was because she felt comfortable. When asked what made her feel comfortable, she straight away referred to the people who cared for each other. All Chinese-Indonesians in Farida's study expressed their comfort in knowing the neighbors, despite the fact that their neighbors were mostly non-Chinese, as well as their feeling of safety during crises. Nevertheless, when asked whether their children continue living in Kampung Peneleh, all of them indicated that their children moved out and chose to live in private housing complexes in the outskirts of the city. The elderly woman revealed that one of her children actually wanted her to move in with them, but she refused.

The pattern of younger Chinese-Indonesians settling outside kampung is also observable in Kampung Pabean, another old kampung located near Surabaya harbor and fish market. In January 2019, I visited the kampung and was introduced to one of the community leaders there who was keen on heritage issues. He took me around Kampung Pabean for a walk while sharing stories about some houses, alleyway corners, the market, and connections with the river. When we passed by an alleyway, one house caught my eye because it had a Chinese mirror attached to the top frame of the main entrance. The community leader offered to facilitate a visit to the house and knocked on the door. An elderly lady came out and greeted us, and she welcomed us into her living room.

The lady was very friendly and offered us a house tour. She showed all the rooms in her house all the way to the kitchen. She had lived there all her life. She told us that, when she was young, she used to make *bakpao* (steamed buns) and sold them for a living. Her younger son came out to join us in the midst of the conversation. Comparable to Farida's findings above, most of the children did not live in the kampung anymore. Although one of her sons still lived with her, this case still corroborates the phenomenon of Chinese-Indonesians moving out of the kampung, particularly the generation that was coming of age and looked for their own place to live during the housing boom.

11.5 The Impact of Discrimination on Segregation and Vice Versa

The young generation of Chinese-Indonesians' tendency to move out of the kampung is inseparable from the systemic discrimination and privatized housing market. The "othering" that takes place through the formal bureaucratic administration results in the construction of the Chinese-Indonesians' distrust of the system. Studies in Medan and Jakarta revealed Chinese-Indonesians' preference to live in gated communities, even when the houses there are more expensive, because they distrust the state in safeguarding their safety after the May 1998 riots (Tsai 2011). Nonetheless, the pattern of Chinese-Indonesian preference to live in private, gated complexes had started long before the riots, and is also observable in Surabaya, where there was no massive anti-Chinese riot in May 1998 (Chong 2018). Chinese-Indonesians who are economically middle-class and above would then be conditioned by this system to look for safety, which was the selling premise of gated communities by private developers. Thus, the need for the sense of safety was also subject to capitalization by the profit-driven housing development. At the same time, Chinese-Indonesians' preference to demarcate their area of residence from the surrounding neighborhoods feeds into the stereotype of the ethnic group as "exclusive and unwilling to mingle" (Kuntjara and Chang 2020, p. 202; Coppel 1983). Development of massive real estate projects and new towns in the periphery, particularly in Jakarta, "had almost nothing to do with spatial planning of the area" (Firman 2004, p. 356), yet they emerged as the "symbol of modernism, security, and lifestyle."

Herlambang et al. (2019) noted a new phenomenon of housing development in Jakarta after the new millennium, which he called "Back to the City." By then, as the suburban housing projects were no longer novel, large-scale developments grew in the city center. However, the problem was that the large-scale housing projects' mission was to encourage people to go back to the city: it would potentially compete for space with the kampungs that were already existing in the city center. Furthermore, these large-scale developments required new infrastructures that would also mean requiring space. "Back to the City" phenomenon emerged after the Reformasi, which had dismantled Soeharto's legacy of the formal discrimination of Chinese-Indonesians. By then, the requirement for citizenship certificate was abolished, and public celebrations of the Chinese New Year returned to the main streets of Chinese quarters across the country particularly during Gus Dur's presidency. Nevertheless, Reformasi brought little change to the housing provision and ethnic segregation that had occurred earlier, facilitated by the housing market, over the decades.

Such segregated living raises potential frictions, especially when relationships among ethnic groups are fragmented. Without first-hand experience that can evoke social and emotional connections, kampung becomes a distant "other" place. Hence, instead of experiencing the kampung as a mixed district in which neighbors know each other, it comes across as a messy and continuously densifying place without sufficient improvement of infrastructures and services. This messiness is in contrast with the neatly planned gated communities. As generations go by, there are fewer Chinese-Indonesians living in various kampung, which leads to the constant "othering" process. A manifestation of such othering is represented in the stereotypes contained in the following excerpt from one of the informants:

> A life-changing moment for me happened during [my high school activity with scavengers], when I went to a slum in the Bermis area. I tried to introduce myself to a child who was running around there, and that child's response to me was shouting: "Cina! Kristen! Kafir!" (Chinese! Christian! Unbeliever! – my translation). That experience shocked me and until today it is still fresh in my mind, even the details of the situation and the face of that child.
>
> (A1, male, 28 years)

The relationship between kampung and urban development is influenced by the city's development challenges. The more intense spatial contestations in a city, the more contradictory the relationship would be. Kampung in Jakarta is probably facing the most stigma. As the city grapples with frequent flooding and worsening traffic congestions, kampung as spaces that contradict the image of the modern city become objects of "improvement," and these "improvements" may include demolition to make way for spaces that are more in line with technical solutions of urban problems. Its narrow alleyways are unsuitable for the growing population of cars, and its land plots are often incompletely translated into well-defined private properties on the map, as the settlements predated the agrarian law. Furthermore, densification of cities also induced relatively newer self-built settlements in which people who could not afford housing from the formal real estate market found their more affordable options, and these settlements often get the blame for making the city messier.

11.6 Anti-Chinese Sentiments?

Therefore, it is of no surprise that kampung also gets the blame for flooding, particularly those located on the riverbanks. As the flooding gets worse and more frequent, riverbank kampung gets more scrutiny, along with images of its inhabitants taking refuge in temporary shelters. However, flooding in Jakarta has added tension to the relationship between Chinese-Indonesianness and kampung. In 2015, Jakarta witnessed its first violent eviction since Reformasi in the name of flood alleviation. The August 2015 eviction of the inhabitants of Kampung Pulo, at the bank of Ciliwung River, involved hundreds of police and military personnel, with the use of violence and tear gas. The eviction took place under the leadership of Jakarta's Governor, Basuki Tjahaja Purnama, a Chinese-Indonesian-Christian who rose to take up the governor position after the previous governor, Joko Widodo, was elected to presidency and had to vacate the seat.

Kampung Pulo (Fig. 11.1) was the first case of violent eviction in Jakarta since the era of Governor Sutiyoso, who earned the nickname "Eviction King"—due to the evictions he had undertaken under his leadership (1997–2007). The atmosphere in the kampung was tense after the eviction of the stretch of houses along the riverbank, especially during the first few weeks.

> The landscape is one of devastation. Rubbles and half-cut homes peppered the scene, even when the "inspection road" along the river was already built. This is not the first time I walk around Kampung Pulo, but this is the first time I sense tension. True enough, as I walked on the "inspection road", I noticed two men looking at us [myself and my colleague, a member of the NGO from across the river]. Eventually, they approached us and asked "What are you doing here?" with an unfriendly tone, clearly one of suspicion. I had never gotten such a tone in my experience entering a kampung for research. [...] When I mentioned that [religious leader]'s name, his face suddenly changed to become more friendly. He then apologized to me and my colleague, saying that their unfriendly welcome was due to the recent eviction, so they were suspicious of any outsider. (Field Note, September 2015).

Flood alleviation projects, such as the concrete embankment of the river, at first glance seems to be consistent with the need to achieve a livable city, one that combines the sectors of environment and infrastructure. Moreover, the demolition of organically grown riverbank settlements that were lacking infrastructure services provided an image of physical "improvement": from one that was messy to one that was neater. Nevertheless, the case of the river embankment project that involved forced eviction

Fig. 11.1 River embankment post-eviction of Kampung Pulo, Jakarta, in late 2015 (Photo by Rita Padawangi 2015)

and escalated ethnic tension illustrates how "liveability" should involve more than the usual checklist to avoid such detrimental effects on the social fabric of the society.

The tension between kampung and Chinese-Indonesianness continued throughout Basuki Tjahaja Purnama's leadership, as kampung evictions continued. LBH Jakarta (Jakarta Legal Aid Foundation) noted that forced eviction cases during his 2-year term were comparable to that of Sutiyoso's 10-year governorship. The adversarial image between a Chinese-Indonesian governor and kampung culminated in Jakarta's gubernatorial election in 2017, which was infused by religious and ethnic sentiment. On the one hand, scholars often attributed Purnama's loss to ethnic discrimination against Chinese-Indonesians as well as religious sentiments, as he was charged with blasphemy for his statement related to a Koranic verse (Padawangi 2023). On the other hand, he was also a governor who had conducted forced evictions of kampung in the name of development, and this contributed to his defeat (Savirani and Aspinall 2017; Gani 2018). All in all, both aspects had added to the adversity between the image of kampung and Chinese-Indonesianness, which is in contrast with the diversity that kampung represents.

Consider the contrast between the case of ethnic tension in Jakarta and the following excerpts from Surabaya, which was based on a field note in 2019:

> A Chinese-Christian RT leader in Kampung Peneleh shared to us that people in Peneleh did not see his identity as an ethnic and religious minority a problem; they valued his good understanding of the kampung. There are also ethnic and religious minorities as heads of neighborhood associations in nearby kampung, such as a Chinese-Christian woman RT leader in Plampitan and a Balinese-Hindu RT leader in Lawang Seketeng. (Perkasa et al. 2022, p. 372)

The case of Kampung Peneleh and Plampitan in Surabaya, which showed relatively less concern about having minority leaders—including Chinese-Christians—in the communities, seems to be in line with the history of the kampung as a mixed district that is in contrast with ethnic zones during the Dutch colonial era. Surabaya also has significant real estate developments in the outskirts of the city, such as CitraLand and Pakuwon City, each with areas more than 500 hectares. However, despite being one of the largest cities in Indonesia, Surabaya still has kampungs as a significant pattern of its urban fabric—social, cultural, and physical built environment—in the center of the city. Therefore, while there are observably hollowing out of Chinese-Indonesians from kampungs throughout the development boom under the New Order (1966–1998), many kampung as spaces of collective governance and collective action still exist, and even their social institutions, such as the Sinoman, still continue (Perkasa et al. 2022).

The adversity between kampung and Chinese-Indonesianness, such as shown in the case of Jakarta, was rooted in the decades-long segregation, indirectly affected by structural discrimination during the New Order era, as well as historic baggage from colonial era that had evolved throughout the years. At the same time, the segregation had also contributed to distancing newer generations of Chinese-Indonesians from kampung as the new real estate development projects dominated housing provision. Adversarial racial sentiments have grown also within Chinese-Indonesians as a perpetuation of segregated social-cultural life, as shown in the following excerpts from Chinese-Indonesian activists:

> Since I was small, **I was unhappy with racist acts and exclusivism of family and Chinese friends towards the non-Chinese**. For example, I was often upset when my friends were angry when being called "Cina" but they themselves uttered the derogatory word "*wana*" to the Javanese, etc. When I was small, I felt [my family and friends] bragged about their superiority: "Tenglang are good with money and have fighting spirit for business. The *wana* are lazy." They were also overprotective and tend to label themselves as victims, although of course they also experienced vulnerability because of their Chinese-ness. (A2, female, 40 years)

> Once, there was a conflict in my mind [when I was in high school] **whether I should hate the "pribumi" people as my family often told me to**, because [the "pribumi"] also planted in their children's minds [to hate the Chinese Indonesians] so there was no use for me to try to change that stereotype. (A1, male, 28 years)

The above narratives reflect the "sense of cultural superiority" among Chinese-Indonesians, which encompasses race and economic class (Kuntjara and Chong 2020). Inter-ethnic stereotyping and negative sentiments are felt and practiced both ways, embedded in multi-generational messages passed down by families. These sentiments are perpetuated by market-driven segregation of places of residence. For these two interviewees, the acknowledgment of those sentiments became the basis to start addressing the stereotypes and racism that were perpetuated through inter-ethnic segregation of social life.

11.7 Kampung Advocacy as Desegregation

All interviewees in this small study were Chinese-Indonesians from various backgrounds, but all of them were involved in kampung advocacy: both to fight evictions as well as to raise awareness on the importance of kampung in envisioning a desired urban life in Indonesia. All of them were chosen on their ability to criticize evictions under Purnama's reign as the governor of Jakarta, at the same time, were also critical of the blasphemy case and ethnic-religious-infused despise of Purnama. The study is qualitative; hence, interviewees were chosen to represent a range of professions and backgrounds: journalist, architect, lawyer, artist, and an employee of a non-governmental organization. There were not so many Chinese-Indonesians involved in the anti-eviction campaign at the time, and therefore documenting their narratives and experiences is important to find out the reasons and conditions that led them to advocate for kampung.

None of the interviewees lived in kampung that were threatened and eventually demolished by eviction, but they feel strongly about the importance of kampung. How do these Chinese-Indonesian activists perceive their relationship with urban kampung? They consistently view their urban kampung activism as acts of desegregation. To them, kampung *kota* represent spatial practices of diversity in everyday life. Their close multi-ethnic interactions in kampung *kota* become the ground of their involvement in urban kampung activism. Such interactions may have started either during their early upbringing or through their immersion into the kampung *kota* life in later years.

> When I was small (in primary school), my family home was in kampung *kota*. I often played with friends from different ethnic groups and religions in the kampung *kota* around the house. (A4, male, 43 years)

Another ground of their involvement in urban kampung activism is their first-hand experience in encountering challenges faced by kampung in the city. Since these challenges vary from one city to another and from one kampung to another, the manifestation of kampung activism may vary from one place to another.

> After the forced eviction of Taman BMW, I became a volunteer with the Humanitarian Network of Volunteers to bring logistic assistance and data collection there. Almost every night, I interviewed kampung residents for data collection. I met a young woman who was heavily pregnant who told me that she had been evicted several times… and also others who said they had been evicted 7–8 times in Jakarta. There was one evening when I heard those stories, suddenly I felt sick and dizzy, I felt very heavy… I could not imagine their heavy burdens. (A5, female, 35 years)

Kampung eviction takes place in many cities across Indonesia; hence, there are networks of solidarity and activism against eviction, such as the Urban Poor Network (JRMK) and *Forum* Kampung *Kota*. The intersection between kampung, eviction, and poverty often makes kampung advocacy intertwine with urban poor advocacy. Kampung activism is not always about anti-eviction; there are various activist groups that focus on service provision and infrastructure improvement as well. Nevertheless, advocacy against kampung eviction is often the most intense, as it carries the agenda to both promote the kampung as a desired form of social life and the right to secure housing tenure. An example of such intense advocacy is the effort of a coalition of activists and residents to prevent the eviction of Bukit Duri inhabitants in Jakarta. After the kampung was evicted in 2016, they filed a class action lawsuit and won. All these took place in urban landscapes that have been shaped through decades of development in which the private sector dominated housing provision, which allows ethnic discrimination to manifest in market-driven housing choices. The perception of gated communities as safer and more desired neighborhoods has led to a segregation of housing choices and the hollowing of kampung from Chinese-Indonesian residents. The dissociation of Chinese-Indonesians from kampung, added with the evictions under Purnama's leadership, had also placed Chinese-Indonesian presence in kampung as something awkward and unsafe, especially in Jakarta, as uttered by one of the activists:

> [After Ahok's forced eviction, researchers asked me] how the kampung residents treated me as a minority Chinese. Whether they treated me differently, whether it was difficult as a Chinese Indonesian to work with kampung residents. They asked whether I was afraid. (A5, female, 35 years)

The stereotype of kampung in Jakarta as a threatening place for Chinese-Indonesians contrasts with the following depiction from a Chinese-Indonesian in Kampung Peneleh in Surabaya:

> Another 72-year-old Chinese-Indonesian interviewee felt a strong bonding with Kampung Peneleh and cited the *gotong royong* spirit that gave the feeling of home. He continues to stay in Gang 3 Peneleh until today (2019), even when the number of Chinese residents has decreased. According to him, the situation in the neighborhood had always been safe for Chinese residents even during the 1965 incident and the 1998 riots. He recalled how residents filled kampung alleyways with cars in 1998 to prevent rioters from entering. (Perkasa et al. 2022, p. 373)

Moreover, kampung advocacy in Jakarta took a toll on the Chinese-Indonesian activists when they faced negative reactions from families and friends. The adversarial relationship between Purnama as governor and kampung led to a widespread inability to differentiate between kampung advocacy and racially driven "anti-Ahok" sentiments; a fragmentation and segmentation that were perpetuated by social media, but have real impact on the everyday social life and practices.

> I was indirectly despised. Friends and family from Chinese descent once labelled me "radical," … when I started to appear in mass media to voice against forced evictions that Ahok as the Jakarta Governor ordered at the time. I also knew that there were a few non-Chinese Ahok supporter friends who turned against me because of my advocacy, and some of them now do not speak to me anymore. (A1, male, 28 years)

11.8 Looking Ahead

Kampung advocacy is inseparable from the threats that kampung is facing amidst rapid urbanization and persistent developmentalism that paints them as antitheses of modern—"improved"—urban living. Kampung advocacy can manifest in many different ways that are also inseparable from the threat that each kampung is facing. The most polarized form of kampung advocacy would be one against eviction, since it is a direct criticism to developmentalism. Kampung advocacy is against the wave of urbanization hegemony and the overarching capitalist urbanization that paints private property as the ideal form of space management.

The uneasy ethnic relations between Chinese-Indonesians and other ethnic groups in Indonesia complicates the relationship between urbanization and kampung. Kampung is historically and essentially a mixed district that reflects how Indonesia as a diverse population can share spaces, keeping their ethnic identities while allowing all residents to participate in the collective governance and collective action to address shared issues. Kampung enables residents, including Chinese-Indonesians as a minority, to identify themselves as members of the same kampung together with residents from other ethnic groups. Yet, market-driven housing provision in Indonesia's urban transition coincides with politicized discrimination of ethnic Chinese during the New Order era and contributed to the segregation of social life, making kampung a less likely choice for the New Order era-generation Chinese-Indonesians even when their parents live in kampung. The dissociation between Chinese-Indonesians and kampung runs counter to the historic meaning of kampung as a place where *bhinneka* (diversity) is part of everyday life, including for ethnic Chinese, but is hardened when kampung is under threat.

The juxtaposition of kampung under development threat, dissociation of Chinese-Indonesians from kampung, and the increased evictions of kampung under an ethnic Chinese governor in Jakarta lead to a convoluted, yet heightened polarization of ethnic tension. Thus, the escalation of ethnic tension during the heated Jakarta elections in 2017 was rooted in the misunderstanding of the intersections of urban life, kampung, and segregation.

This small chapter, which documents the experiences of Chinese-Indonesian kampung advocates alongside the challenges of development and historic diversity of kampung, is a contribution to present an aspect of livability that is crucial, yet rarely, if ever, discussed in the indices. Kampung is a space in which collective governance and collective action are part of everyday life, and one that becomes the basis for Chinese-Indonesian urban activists to advocate against segregation and against forced evictions. As this chapter demonstrated, kampung is a spatial manifestation of diversity that becomes more intense and complex as Indonesia's New Order urban transition makes it denser than ever. Yet, in kampung, we can find hints of how Indonesia's diverse population shares space and expresses abilities for collective organization of space, which becomes the basic necessity to shape a civil society that acts both for and beyond itself. Practices of tolerance that start from kampung is imperfect; there are power inequalities as well in the local dynamics, but no matter how imperfect it may be, there is a sense of human agency that is involved, as all residents of the kampung can have a say in terms of the organization of space. The messiness of the kampung is the manifestation of that human agency and collective organization.

The hollowing of Chinese-Indonesians from kampung, however, weakens the share of ethnic Chinese from spatial manifestations of collective organization. As gated real estate projects become more desired, the segregation that ensues makes the city more vulnerable to ethnic tensions. While some of the older Chinese-Indonesians in Kampung Peneleh expressed their appreciation of kampung as a safe and comfortable place, the image of kampung as a place that is less developed often results in eviction threats in the city.

The documentation of Chinese-Indonesian kampung advocates-activists in this chapter is an effort to bring back the nuances of diversity and the importance of spatial perspectives in seeing ethnic sentiments and tolerance in an increasingly diverse urban Indonesia. The hardening of ethnic sentiments has been affected by ways of seeing the issue, while in fact the problem is rooted in developmentalism that threatens kampung as the historic *bhinneka* settlements of urbanizing Indonesia. Urban kampung advocacies are spaces in which Chinese-Indonesian activists form relationships and develop spatial rootedness in the neighborhoods. As their primary residences are rarely in urban kampungs, the potential of desegregation remains limited, but interactions through urban kampung activism lead to a re-negotiation of the sense of Chinese-Indonesianness.

> Now I can safely conclude that for many people who have a certain idea about Chinese-Indonesians, I am not "Cina" enough. But instead of making them shed their stereotyping on Chinese-Indonesians, my atypical behavior only makes them regard me as "unChinese". I am a different kind of "Cina". But I'm not doing this to debunk the stereotyping anyway, so

even though I am a bit disappointed that my existence does not make them think differently about Chinese Indonesians, I won't stop doing the kind of [work] that I believe. (A3, female, 43 years)

References

Chong, W.-L. 2018. *Chinese Indonesians in post-Suharto Indonesia: Democratisation and ethnic minorities.* Beijing: Hong Kong University Press.

Coppel, C. A. 1983. *Indonesian Chinese in Crisis.* Oxford: Oxford University Press.

Farida, E. N. 2019. *Screams, tears, and laughters: Retracking Chinese-Indonesian memory over spatial and cultural transformation in* Kampung *Peneleh, Surabaya.* Paper presentation. Chinese Indonesians: Identities and Histories, Monash Herb Feith Centre Conference 2019.

Firman, T. 2004. New town development in Jakarta Metropolitan Region: A perspective of spatial segregation. *Habitat International* 28 (3): 349–368. https://doi.org/10.1016/S0197-3975(03)00037-7.

Gani, E. 2018. *Answering Jakarta: My step to move on.* Medium. https://medium.com/@edbert.gani/answering-jakarta-my-step-to-move-on-a27e91b48785

Herlambang, S., H. Leitner, L. J. Tjung, E. Sheppard, and D. Anguelov. 2019. Jakarta's great land transformation: Hybrid neoliberalisation and informality. *Urban Studies* 56 (4): 627–648. https://doi.org/10.1177/0042098018756556.

Kim, J.-H. 2002. Research notes on the making of a "Gated Community": A study of an inner city neighbourhood, Jakarta, Indonesia. *Asian Journal of Social Science 30*(1), 97–108. https://www.jstor.org/stable/23654626

Kuntjara, E., and C.-Y. Hoon. 2020. Reassessing Chinese Indonesian stereotypes: Two decades after Reformasi. *South East Asia Research* 28 (2): 199–216. https://doi.org/10.1080/0967828X.2020.1729664.

Kuntjoro-Jakti, D. (1986) *Kemiskinan di Indonesia* [Poverty in Indonesia]. Yayasan Obor Indonesia.

Kusno, A. 2014. *After the new order: Space, politics, and Jakarta.* London: University of Hawaii Press.

Padawangi, R. 2022. *Urban Development in Southeast Asia.* Cambridge: Cambridge University Press.

Padawangi, R. 2023. Spatial perspectives: Civil society activism as struggles for space in urbanising Southeast Asia. In: Hansson, E. and M. Weiss, editors. *Routledge handbook of civil and uncivil societies in Southeast Asia.* 65–80. https://doi.org/10.4324/9780367422080-5

Padawangi, R., and M. Douglass. 2015. Water, water everywhere: Toward participatory solutions to chronic urban flooding in Jakarta. *Pacific Affairs* 88 (3): 517–550. https://doi.org/10.5509/2015883517.

Perkasa, A., R. Padawangi, and E. N. Farida. 2022. The kampung, the city and the nation: Bhinneka Tunggal Ika in the everyday urban life of Kampung Peneleh, Surabaya, Indonesia. *Asia Pacific Viewpoint* 63 (3): 364–378. https://doi.org/10.1111/apv.12359.

Savirani, A., and E. Aspinall. 2017. Adversarial linkages: The urban poor and electoral politics in Jakarta. *Journal of Current Southeast Asian Affairs* 36 (3): 3–34. https://doi.org/10.1177/186810341703600301.

Silas, J. 1996. *Kampung Surabaya menuju metropolitan, Surabaya.* London: Yayasan Keluarga Bhakti.

Tsai, Y.-L. 2011. Spaces of exclusion, walls of intimacy: Rethinking "Chinese exclusivity" in Indonesia. *Indonesia* 92: 125–156. https://doi.org/10.5728/indonesia.92.0125.

Part III

Chapter 12
Epilogue Toward Livable Kampungs

Melani Budianta

Abstract The epilog highlights the main thread of the edited volume, reflecting on the sustainability of Indonesian kampungs in the threat of urban expansion. Contextualizing the kampungs within "middling urbanism" (Kusno, 2023), the epilog argues that as a mediating space between the urban and the rural, kampungs have important role in supporting the urban economy, a fact that often overlooked in urban planning. In negotiating with the forces of urban expansion, the kampungs in Jakarta and in Nangapanda, East Nusa Tenggara, deal with diverse socio-environmental challenges through commoning initiatives to keep their kampungs viable and liveable.

keywords Kampung · Sustainability · Mediating space · Negotiation · Communing

By the end of 2023, the rate of urbanization in Indonesia has reached 58.57% from 51.96% in 2013 (O'Neill 2024). Land use change from the conversion of rural agricultural field into urban amenities have resulted in the loss of agricultural fields from 150,000 to 200,000 hectares per year. In Java island this means the loss of harvested rice up to 3 million (PakTaniDigital 2021).

Urban expansion has also changed the spatial and physical environment of the villages, from rural to urban villages or *desa perkotaan* in the classification of state regulation. The growth of urban villages in Indonesia nearly doubled from 2010 to 2020, from 15,786 to 29,640. Compared to rural villages, these kampungs have denser population, higher ratio of residential area to farming areas, provision of electricity and cable telephone system, and urban facilities like school, shop, hospital, and amenities like hotel, discotheque, massage parlor, and hair dressing salon (Badan Pusat Statistik Indonesia 2020).

In line with this tendency, Thompson (2021) argues that the rural villages experiences "urban annexation". He examines how "kebun culture"—which is initially

M. Budianta (✉)
Universitas Indonesia, Depok, Indonesia
e-mail: melani.budianta@gmail.com

M. Budianta et al. (eds.), *A Liveable Kampung*, Engaging Indonesia,
https://doi.org/10.1007/978-981-96-7866-2_12

associated with the rural cultivation of nature is in the context of Malay peninsula—has been coopted by the urban actors to enact a nostalgic return to the rural villages. Thompson uses the term "urban annexation", because reverse migration from urban to the rural area brings capital home to carve orchid gardens which are detached from the social lives of the villagers. The garden serves as a resort unconnected to the surrounding agricultural environment. The *kebun* culture also reinforces social gap and power relations between capital owner and rural workers.

In another neighboring country, with fast-paced modernization like Singapore, the word kampung or village only remains as a nostalgic reminder to a past in an area that has been gentrified for the middle classes. With this urbanizing trajectory, will the rural in Indonesia and Southeast Asia be in the process of being erased, absorbed, and changed into urban environment?

Abidin Kusno (2023) finds this scenario of urban expansion as "insufficiently nuanced", and argues instead for a "middling urbanism"—in which capitalist modernization "incompletely urbanizes" the city (967). The focus of Abidin Kusno's research is the urban kampung in Jakarta, informal settlements in the peripheries of the megacity. The kampung is connected to the rural—as it is where the rural migrants reside to make a living in the city. Through the "cut in the wall" the kampungs are connected to the gated residential areas and city centers. As an affordable space to supply the workforce of the city, and a space where community spirit is still alive, the urban kampung and the city creates a symbiosis. With the interplay between the capital and irregular state regulations, Kusno claims that "despite the claims of urban transformation" the kampung "continues to exist and grow in many different forms" (p. 957).

This edited volume underlines the fact that urban expansion is still occurring in fast mode all over Indonesia. At the same time, while acknowledging the precarity of urban and rural kampungs in the face of the capital and structural power relations, the scholars in this edited volume affirm the social and economic value of kampungs and their tenacity in negotiating with, and at times riding on the waves of urban expansion. The focus of this book is on the livability of kampung and the inhabitants in fulfilling their daily needs and wellbeing—as seen from various angles and dimension—nutrition and health, environment, and socio-cultural aspects.

How livable are kampungs in Indonesia, rural as well as urban, and how do they fare in the urbanized and globalized future? This is the question that fuels the chapters in this volume. Nine out of the twelve chapters are the results of a multidisciplinary and transdisciplinary research involving researchers from the fields of social-humanities, health sciences, economics, and psychology of the Universitas Indonesia in 5 urban kampungs in Greater Jakarta area, and 33 rural villages in Nangapanda Sub District, Ende, Flores.

Kampung in this edited volume refers to informal settlements in the urban as well as rural areas. In the rural areas, they are usually referred to as villages and have specific geographical as well as administrative boundaries. In urban areas, kampungs are self-identified as a community of place that might not be identified similarly in the city public administration.

The two sites—in the Greater Jakarta area and the eastern part of Indonesian archipelago—reflect different scales of urban expansion and their proximity to Jakarta, the capital as well as the epitome of the most urbanized space in the country. Yet, even the remote villages of Nangapanda have been urbanized, if not physically, in their consumption and lifestyle.

Within the greater Jakarta area—abbreviated as Jabodetabek (Jakarta, Bogor, Depok, Tangerang, Bekasi), the five kampungs spread in different locations with their own specific demographic features as well as conditions of urbanity and its problems. The kampungs offered are located in different area across the Greater Jakarta area with different issues of livability. One kampung, called Gedung Pompa, is at the Jakarta bay to the north of Jakarta, with issues of clean water and the rise of sea level. Kampung Nambo in the southwest, on the other hand, is abound with natural water from the valley springs, yet the health and wellbeing of the kampung dwellers are being challenged by the landfill that the local government dumped on the hill across their kampung. The urban kampung of Cikini in the heart of Jakarta has much greater access to facilities than the other kampungs. Yet, the inhabitants face daily challenges of facing drug dealers and pollution. Kampung Cimone, and Markisa, located behind the newly grown industrial estate in Tangerang, the sister city of Jakarta, serve as the homes of the industrial unskilled workforce. The two kampungs try their best to deck the narrow alleys and corners in their kampungs to create a livable public space, among other things by doing urban farming. With limited income, many kampung dwellers in the five different places keep their stamina by buying cheap cure-all unregistered medicine or make their own herbal medicine.

The five urban kampungs in Jakarta, however diverse their contexts, could not represent the 82,395 villages all over Indonesia, in particular those located outside Java. The further away to the east, the less these villages have access to urban facilities. With the exception of Kampung Nambo, the four other kampungs in Urban Jakarta face the issue of crowding and space, while space is not an issue in the Nangapanda villages. Air quality, which is the biggest hazard in Kampung Nambo, and to a lesser degree in the other urban kampungs in the Greater Jakarta area, is not experienced by inhabitants of the Nangapanda villages. However, all kampungs, including in Nangapanda, albeit considered by some writers as still within acceptable standards—have varying degrees of problems with access to good quality of water, sanitation, which affect hygiene and health.

The six chapters on the Jakarta kampungs have shown that in order to fulfill their basic needs, the residents must muster creative efforts and strategies. Residents of urban kampungs generally spend a good portion of their income in getting clean water for drinking or for household use, and also in arranging for waste management. Kampungs are not generally served with a treated water system by their local government, thus they must obtain their own clean water for drinking as well as for household needs. In areas like Gedung Pompa, which is near the bay, water service providers fill their need for clean water for drinking as well as for household use.

In order to take care of their own health kampung residents either use traditional herbal medicine, or unprescribed cure-all medicine called “obat rombongan”. “The mixed medicines consisted of at least four medicines: antibiotics, analgesics, corticosteroids, antihistamines, antidiarrheals, and vitamins". Based on the analysis made in the two related chapters on herbal medicine and self-medication, the consumption of self-medication by kampung inhabitants occurs in the following pattern: "The provisional conclusion is that the closer a village is to the site of modernity, the lesser the trust in the effectiveness of traditional medicine. It can be seen from Kampung Cikini and Gedong Pompa which are in ring 1 of modernization". On the other hand, inhabitants in Gedung Pompa are likely to consume the unprescribed and unmonitored *obat rombongan.*

The lack of adequate public facilities to fulfill the basic needs of kampung inhabitants has prompted initiatives for collective collaboration between residents, as well as between residents with external entities such as NGOs and kampung activists. The social interactions vary in each kampung, depending on the social relations between older residents and newer migrants or those who rent kampung houses or rooms for temporary stay. However, the commoning initiatives to increase the livability of their kampungs, be it in creating a public toilet, organizing clean up routine of public spaces—or in developing a close-knit atmosphere by making events to celebrate public holidays—do not guarantee sustainability. One looming threat in the kampung are evictions by the local government for the sake of gentrification, or urban development.

One crucial aspect that is often ignored in the government's plan of urban development is the economic and social functions of kampung in the expanding megapolitan like Jakarta. Kampung Cikini Kramat, in lieu of its proximity to the Jakarta urban centers, serves as spaces for lower middle-class workers and clerks to find cheaper accommodation and to get affordable lunch and snacks provided by the kampung residents. Similarly, Kampung Nambo, which has its own water spring, used to be the provider of clean water for its surrounding residential areas. Apparently, this symbiosis between kampung settlements and the rest of the city has not been considered in the planning of the city, as is seen in the fate of Kampung Nambo.

If urban kampungs within and in the Greater Jakarta area shows precarity in their livability—primarily in accessing basic needs, what about kampungs in rural areas in Eastern islands? Although Indonesia after the decentralized policy of 2001 have given local government authority in governing their regions, social and economic capital has not been equally distributed—especially in the Eastern part of Indonesia. The three chapters in the second section support this general assumption. One apparent issue is provision of adequate educational facilities for children at primary schools, the problems in accessing clean water in the uphill areas, and the high percentage of microbes in the household drinking water, which causes various health problems. Most telling evidence of the kampung predicaments in Nangapanda is the prevalence of stunting and impaired growth in children of the Nangapanda villages, which surprisingly is not due to poverty. Ironically, many rural villages in East Nusa Tenggara reap better income by converting from subsistence farming to tending on cash crop trees such as cacao, rubber, oil-palm, and coconut. Interestingly, the chapter about family eating habits shows that stunting occurs, even when families know how to put good nutrition on the dinner table and even though parents profess that they have regular family meals. The chapter notes that the pre and post meal consumption of children are not monitored. In my own brief visit to Nangapanda villages, I often met children eating packaged snacks and sweets sold in the snack/cigarette stalls in the village. In villages which have shifted from small fields to cash crops farming, the garden—which provides healthy vegetables—seems likely to be neglected, and family meal gatherings less than before due to the long hours of work. Further research on children's eating habit needs to be done, also in families, whose breadwinners are migrant workers. The chapter on the relation between stunting and cognitive stimulations indicate that lower education of parents contributes to parents' lack of confidence in educating and transmitting their knowledge to their children.

The last chapter of this edited volume focuses on one aspect that is mentioned in Chapter 2 about the history of kampungs in Jakarta, but not elaborated further in the rest of the book, i.e., the racial dimension of urban kampungs in Indonesia and its implications on social relations and kampung activism. Writers of Chapter 2 have indicated commoning initiatives done internally among kampung residents and external collaboration between kampung dwellers and organizations as well as activists. The last chapter revisits the two topics by extending the discussions to other kampungs in Jakarta, Surabaya (East Java) and Medan (North Sumatera).

In one of the oldest urban kampung in Surabaya, Kampung Peneleh, the writer, Rita Padawangi, met Chinese inhabitants residing in a multicultural environment. She then discusses the demographic changes in urban cities with the development of real estates for middle and upper middle classes, leaving the sprawling informal kampungs for rural migrants and lower classes workers and creating racial demarcation—as the Chinese moved out of kampungs to the developed areas of the cities. There are different historical contexts of the spatial and racial segregation in the kampungs in Jakarta, Medan, and Surabaya. The chapter ends with discussion about the return of Chinese kampung activists, who were called upon by their childhood memories in urban kampungs to join the commoning initiatives to enhance kampung livability.

The focus on collaboration and commoning is indeed relevant in evaluating the sustainability and livability of the Indonesian kampungs. These chapters show how kampung dwellers work collectively to gather resources that they have to be organized, developed, and shared for the wellbeing of the community. Theorized by Elenor Ostrom in her argument against Garret Hardin's "tragedy of the commons", the collective organizing of shared public facilities or goods has been termed by recent scholarship as commoning initiatives (Fournier 2013). In the wake of global crisis and pandemics in 2020–2022, commoning has become the means for kampung communities to overcome health, economic, and social problems. The need to garner communal effort in time of crisis also indicate the failures of the state, which operates within the neoliberal, capitalistic regime, to provide the communities for their basic need, facilities and welfare. In fact, the revenue-based orientation of local or central government often looms large as impending threat of eviction. In rural and forest areas with rich mineral resources, extractive industries sponsored by the state and the capital have cost the indigenous communities their kampung and natural environment.

Recent trends in kampung activism shows the rise of commoning initiatives as alternatives to exploitative privatization by engaging residents as well as local governments, NGOs, and private companies as stakeholders. The initiative of the South Tangerang City district officer, Suli Rosidy, to engage 30 most unhealthy kampungs in Jakarta to join the Clean and Healthy Living Program in 2013 (discussed in the Introduction chapter) might be a top-down decision, which in the end result in a collective awakening. Among one kampung that made it to the final list, was Kampung Markisa, one of the five kampungs discussed in the early chapters. The name, Markisa, is an abbreviation of "Mari Kita Sadar" (Let Us Change)—symbolized by the *markisa* fruit (passion fruit) planted along their alleys. This new name was adopted to show their collective determination to transform the kampung space to be a livable and pleasant place to stay.

In today's fast pace urban expansion, will urban and rural kampung be there in the future? Urbanism is an unavoidable trend, but the chapters in this edited volume speak of various possibilities, be it in the middling urbanism or a healthier and robust rural modernity. The multidisciplinary approach to kampung speaks of the spirit for collaboration, in the realms of knowledge commoning for livable communities in the future.

References

Badan Pusat Statistik Indonesia. 2021. *Peraturan Kepala BPS no 120 Tahun 2020 tentang Klasifikasi Desa Perkotaan dan Perdesaan di Indonesia* (Regulation of the Head of Central Bureau of Statistics no 120 Year 2020 about The Classification of Urban Villages and Rural Villages in Indonesia 2020. https://www.bps.go.id/id/publication/2021/05/26/cff43de20a058e9e8400ca57/peraturan-kepala-badan-pusat-statistik-nomor-120-tahun-2020-tentang-klasifikasi-desa-perkotaan-dan-perdesaan-di-indonesia-2020---buku-2-jawa.html

Fournier, V. 2013. Commoning: on the social organisation of the commons. *M@ n@gement* 16(4), 433–453. https://management-aims.com/index.php/mgmt/article/view/4020

Kusno, A. 2023. Middling urbanism: The megacity and the kampung. In *Changing Asian urban geographies: Urbanism and peripheral areas*, ed. F. Wu and R. Keil, 51–67. London: Routledge. https://doi.org/10.4324/9781003299981.

O'Neill, A. 2024. *Indonesia: Urbanization from 2013 to 2023*. Statista. https://www.statista.com/statistics/455835/urbanization-in-indonesia/

PakTaniDigital. 2021. *Penyusutan Lahan Pertanian setiap Tahunnya*. https://paktanidigital.com/artikel/penyusutan-lahan-pertanian-setiap-tahunnya/#google_vignette

Thompson, E. C. 2020. Urban annexation of the rural: Kebun culture in Malaysia. *Singapore Journal of Tropical Geography* 41 (1): 136–153. https://doi.org/10.1111/sjtg.12301.

MIX
Papier aus verantwortungsvollen Quellen
Paper from responsible sources
FSC® C105338

If you have any concerns about our products,
you can contact us on
ProductSafety@springernature.com

In case Publisher is established outside the EU,
the EU authorized representative is:
Springer Nature Customer Service Center GmbH
Europaplatz 3, 69115 Heidelberg, Germany

Printed by Libri Plureos GmbH
in Hamburg, Germany